Thin Plate Design For Transverse Loading

The Constrado Monographs deal with the application of
steel in construction. They each treat a specific subject,
and the texts are written with authority and expedition.
Subjects are treated in depth and are taken to the point
of practical application.

Advisory Editor
M. R. HORNE, MA, ScD, FICE, FIStructE
Professor of Civil Engineering, University of Manchester

Already published
The Stressed Skin Design of Steel Buildings
by E. R. BRYAN, MSc, PhD, FICE, FIStructE
Professor of Structural Engineering, University of Salford

Composite Structures of Steel and Concrete
Volume 1. Beams, Columns, Frames and Applications in Building
by R. P. JOHNSON, MA, MICE, FIStructE
Professor of Civil Engineering, University of Warwick

CONSTRADO MONOGRAPHS
Thin Plate Design For Transverse Loading

B. AALAMI, BSc, DIC, PhD, CEng, MICE, MRINA, MASCE
Associate Professor of Structural Mechanics,
Arya-Mehr University of Technology, Tehran, Iran
D. G. WILLIAMS, BE, MS, DIC, PhD
Project Engineer,
Redpath Dorman Long, Bedford, England

Crosby Lockwood Staples   London

Granada Publishing Limited
First published in Great Britain 1975 by Crosby Lockwood Staples
Frogmore St Albans Herts and 3 Upper James Street London W1R 4BP

Copyright © 1975 Constructional Steel Research and Development Organisation

All rights reserved. No part of this publication
may be reproduced, stored in a retrieval system,
or transmitted, in any form or by any means,
electronic, mechanical, photocopying, recording,
or otherwise, without the prior permission of
the publishers

ISBN 0 258 96991 1

Printed in Great Britain by
William Clowes & Sons, Limited
London, Beccles and Colchester

Preface

The limitations of the linear, small-deflection theory of plates have long been appreciated, but it is only recently that computational procedures have made it feasible to contemplate using non-linear theory in design. Even with recent improvements, the engineer will only wish to apply non-linear theory when it appears essential in order to approximate realistically to the stiffness and strength of his structure. How is he, without expensive exploratory finite element or finite difference computations, to decide when he should abandon linear theory, and is there any way in which he can obtain a reliable (even if approximate) estimate of the non-linear solution to his problem? Such considerations can arise in all forms of container vessels using flat plates, in ships and steel bridges, and wherever stiffened or unstiffened flat plates are used to take local lateral load or as main structural elements.

It is planned to publish, within this Constrado Monograph Series, a series of volumes which will provide the designer with a comprehensive review of results obtained from the application of elastic large deflection theory to the analysis of thin plates. This first volume deals with plates under transverse loads, and the second will deal with plates under in-plane loading (post-buckling behaviour). The possibility is also considered of publishing a third volume which will deal with special problems of importance, including those concerned with combined loading.

In the present volume, the aim is to present the solutions in such a form that reliable approximate estimates can also be made of maximum stresses and deflections in cases not covered by the charts and tables. To make this possible, the results have been expressed in terms of coefficients applied to the linear solutions, so that, if the linear solution to any other problem is available, it becomes easier to obtain an estimate of the modification due to non-linear effects.

To obtain full advantage from the data in this volume, a proper understanding of the physical phenomena involved is essential. Some appreciation of the mathematical basis is also desirable, although not essential. The authors present a clear summary of the theory involved, and complete stress distributions for a number of typical cases. Careful study of these latter is recommended, since they contain important lessons for the proper understanding of the systematic tables. Finally, the use of the charts and tables for typical design situations is discussed and illustrated by examples.

It need hardly be said that the computational effort represented by this volume is considerable. The user should not expect to extract his information without some effort of understanding of the principles on which the volume is based. Given that effort, the monograph should undoubtedly open the way for the more soundly based design of transversely loaded plate structures than has hitherto, except by considerable computational effort, been possible.

Both authors have been actively engaged over the last decade on research into thin plate structures, both in experimental work and in the development and use of computational procedures for thin plate structures.

M. R. Horne
August 1974

Contents

Preface   v
Foreword   xi
Notation   xv
Units and Conversion Factors   xvii

Chapter 1. General Considerations
1.1 Behaviour of Thin Plates   1
1.2 Large-deflection Behaviour   2
1.3 Historical Background   5
1.4 Basic Plate Relationships and Equations   6
    1.4.1 Sectional Actions
    1.4.2 Stresses on an Element of Plate
    1.4.3 Basic Plate Relationships
    1.4.4 Governing Plate Equations
    1.4.5 Boundary Conditions
    1.4.6 Outline of the Solution Procedure

Chapter 2. Large-deflection Design
2.1 Large-deflection Design Considerations for Plates Under Transverse Loading   18

viii Contents

2.2 Geometry, Boundary Conditions and Loading 18
    2.2.1 Geometry
    2.2.2 Boundary Conditions

2.3 Selection of Design Criteria 31
    2.3.1 Control of Deformations
    2.3.2 Control of First Surface Yield or Initiation of Fracture
    2.3.3 Allowance for Limited Spread of Yield Through the Thickness
    2.3.4 Control of Mid-plane Yielding of Plate

Chapter 3. Design Parameters, Design Procedure, and Description of Typical Plate Problems Presented

3.1 Design Parameters 37
3.2 Design Procedure 38
3.3 Poisson's Ratio 40
3.4 List of Plate Problems Treated 41

Chapter 4. Design Curves and Data

4.1 Symmetrical Plates Under Uniform Transverse Loading 44
    4.1.1 Design Curves
    4.1.2 Detailed Description of Large-deflection Behaviour
    4.1.3 Approximate Solutions and Plates with Infinite Aspect Ratios
    4.1.4 Related Work by Other Investigators

4.2 Rectangular Plates Under Central Concentrated Loadings 69
    4.2.1 Design Curves
    4.2.2 Related Work

4.3 Rectangular Plates with Unsymmetrical Boundary Conditions Under Uniform Transverse Pressure 84
    4.3.1 Design Curves

4.4 Rectangular Plates Under a Linearly Varying Distribution of Transverse Pressure (Hydrostatic Pressure) 94
    4.4.1 Design Curves
    4.4.2 Edge Beams
    4.4.3 Related Work

Contents ix

    4.5   Circular Plates Under Uniform Transverse Pressure or Concentrated Loadings   104
          4.5.1 Design Curves
          4.5.2 Related Work

Chapter 5. Numerical Examples

Example 5.1.	Steel Plates Under Uniform Transverse Pressure	109
Example 5.2.	Riveted Aluminium Plate Under Transverse Loading	111
Example 5.3.	Aspect Ratios Not Specifically Covered—Corner Panels	113
Example 5.4.	Glass Window of a High-rise Building—Variations in Poisson's Ratio	115
Example 5.5.	Steel Plate Under Concentrated Loading	117
Example 5.6.	Side Wall of a Liquid Container Resting on an Edge Beam at Top	119
Example 5.7.	Penetration of Yielding Through the Thickness	121

References and Bibliography   123

Appendix. Data Tables for Large-deflection Elastic Behaviour of Transversely Loaded Plates

A1	Notation	129
A2	General Description	130
A3	Boundary Conditions	131
A4	Loading	132
A5	Small-deflection Data—Linear Coefficients K_l	134
A6	Large-deflection Data—Reduction Coefficients K_r	136
A7	Shearing Stresses τ_b, τ_m	139
A8	Signs	139
A9	Poisson's Ratio	139
A10	Summary of Some Useful Relationships for Use with Data Tables	140
A11	Additional Numerical Examples	142

Index   193

Foreword

Thin-plated structures have gained special importance and notably increased application in recent years. Ships, steel bridges (in particular, box girders), plate girders, containers and aircraft are a few of the major large-scale structures using metal plates. Thin plates also frequently form a part of machinery components and other devices. New advances in development and use of synthetic materials as principal load-bearing components have added a new dimension to the application of thin plates as structural elements.

Significant savings in weight may be achieved if thin plates are designed with due consideration to the large-deflection elastic pre- and/or post-buckling behaviour. Moreover, large-deflection analysis, being capable of a more exact representation of plate behaviour than the classical theory of plates, leads to a more rational and safer design.

It is only through recent advances in numerical and computational techniques that the large-deflection problem of elastic plates can be treated satisfactorily. Developments of large-deflection theory to its fullest potential has transformed the theory from a research subject into an applied and promising design tool. The growing demand for its application in design is due to economic and safety considerations, as well as recent structural failures resulting from a direct lack of the appropriate design knowledge. As a matter of fact, large-deflection analysis has been used successfully as a design tool in several recent ship and bridge structures, where the need for sophistication in design, economical competitiveness and yet safety of the structure made the use of more exact methods of design necessary.

New design codes for steel structures being drafted along the lines of the Merrison Committee Report, which recommends large-deflection theory as a design method, create a strong need for immediate provision of the relevant design knowledge and data. The book aims to familiarise the design engineer with the physical and analytical concepts of the large-deflection behaviour of plates and to provide him with the necessary design tables, design figures and simplified formulae in order to enable him to aim at a more rational and weight-saving design. The book will hence be of interest to stress analysts in the field of plated structures, structural engineers, researchers in the field of plate theory and elasticity and also graduate and post-graduate engineering students with a mind for more accurate, recent and rational methods of analysis and design.

This book deals with the introduction to large-deflection behaviour and design, with particular reference to plates under transverse loading. It contains a comprehensive collection of design tables and graphs of over 100 cases covering the more common cases of laterally loaded elastic plates, together with ample design examples for their interpretation and use.

The many design curves and tables presented in this volume are based on original calculations. Where data is extracted from the work of other investigators, due credit is given to the source.

The arduous and complex task of carrying out the necessary computations for this volume would not have been possible without the help of Messrs A. A. E. Bayati and E. Shayan for their efficient computer programming and the skilful running of the individual cases. The development of the necessary computer programs and their subsequent use were all carried out using the facilities of the Arya-Mehr University of Technology Computer Centre, for which the authors are grateful.

The authors are indebted to Professor M. R. Horne for helpful discussions, which resulted in shaping the volumes in their present form, being compact and easy to use with a strong emphasis on practical applications. The authors wish to express their gratitude to Dr J. C. Chapman for his initial guidance at Imperial College in the development of the subject, and for his constant encouragement and many valuable discussions on the material treated in this book.

Notation

a	plate side in x-direction
b	plate side in y-direction
D	$Et^3/12(1-v^2)$
E	Young's modulus of elasticity
FBC	flexural boundary conditions
f	Airy's stress function
I	edge beam moment of inertia
M_x, M_y, M_{xy}	moments per unit length of plate
MBC	membrane boundary conditions
N_x, N_y, N_{xy}	membrane forces per unit length of plate
P	$ap/t^4 E$
p	total magnitude of applied patch loading
Q	$a^4 q/t^4 E$
Q_x, Q_y	shearing forces in z-direction
q	intensity of transverse loading
t	plate thickness
UDL	uniformly distributed loading
u	length of patch loading in x-direction; mid-plane displacement in x-direction
V_{max}	maximum value of reaction of edge beam

Notation

V_x	reaction on edge beam at distance x
v	length of patch loading in y-direction; mid-plane displacement in y-direction
W	w/t
w	transverse deflection of plate
x, y, z	rectangular coordinates
α	u/a; coefficient of overstress at plate surface
β	v/a; coefficient for penetration of plastic zone through plate thickness
γ	shearing strains
$\varepsilon_x, \varepsilon_y$	direct strains
ν	Poisson's ratio
ρ	b/a, aspect ratio
σ_x, σ_y	stresses in x- and y-directions
σ_y	yield stress
$\bar{\sigma}$	$(a/t)^2 \sigma/E$
τ	shear stress

Superscripts

$(\bar{})$	non-dimensional stresses given by $\bar{\sigma} = (a/t)^2 . \sigma/E$
$(*)$	quantities referring to Poisson's ratio $\overset{*}{\nu}$

Subscripts

$1, 2, 3, \ldots$	refer to locations on plate
a	average
b	refers to bending action
e	equivalent stress
m	refers to membrane action
max	maximum
x, y, z	coordinate directions
$(,)$	represents partial differentiation in turn with respect to each subscript variable following

Units and Conversion Factors

The design curves and tables presented in this volume are all dimensionless and may be used directly in any system of units. However, the numerical examples given are worked out in SI units followed by their English equivalents in parenthesis. The following conversion table lists the major conversion factors used.

To convert	to	multiply by
inches (in)	millimetres (mm)	25·40
millimetres (mm)	inches (in)	0·03937
pound force (lbf)	newtons (N)	4·45
newtons (N)	pound force (lbf)	0·2247
kilogram force (kgf)	newtons (N)	9·806
newtons (N)	kilogram force (kgf)	0·1020
pounds per square inch (psi)	newtons per square millimetre (N/mm^2)	0·006895
newtons per square millimetre (N/mm^2)	pounds per square inch (psi)	145·0
kilogram force per square centimetre (kgf/cm^2)	newtons per square millimetre (N/mm^2)	0·09806
newtons per square millimetre (N/mm^2)	kilogram force per square centimetre (kgf/cm^2)	10·20

CHAPTER ONE
General Considerations

1.1 Behaviour of thin plates

Metal plate is one of the major components of many heavy structures such as ships, steel bridges, aircraft, sea platforms, etc. Non-metallic plate, such as sheet glass and plywood, also has a wide range of application in lighter structures. Through the use of newly developed high strength synthetic materials it is anticipated that thinner and lighter plates will find increasing application in both heavy and light structures, thus posing a greater demand for more rational and economic methods of design.

Most thin plate components of plated structures undergo deflections ranging from about 0·1 to several times the plate thickness under working conditions. Due to these finite out-of-plate deflections w, the stiffness of the plate changes, resulting in greater resistance to loading, not predicted by classical plate theory. In most cases deformation is elastic, although there are certain applications where a limited amount of plasticity is acceptable under working conditions.

In some recent work (Aalami 1972a,c) it is shown that the usual procedure of plate design, which may be called small-deflection design and is based on classical theory of plates, can overestimate the deflections and the stresses of thin plates by up to 100% for a number of practical problems. Many designs are therefore unduly conservative, and design methods are required which allow for the stiffening effects due to the stresses which develop as a result of finite deflections.

The usual function of a thin-plate element is to withstand a lateral loading, or to act with the adjoining structure in sustaining in-plane forces, or both. This book deals with plates under transverse loading alone. In the context of this book a thin plate is one which, under loads, changes its load-carrying characteristics due to its out-of-plane deformation w by enough to warrant the consideration of this change in order to achieve an economic design. Such a plate is said to have undergone large deflections, and design for these deformations will be referred to as large-deflection design.

For the purpose of large-deflection design it is not sufficient to define a thin plate simply by its breadth-to-thickness ratio (b/t). The magnitude of the out-of-plane deflection is governed by several factors as outlined below, in addition to plate geometry.

(1) Magnitude of transverse pressure Q

A plate's breadth and thickness, together with the transverse pressure and the Young's modulus of elasticity, form the non-dimensional loading parameter Q defined as $(a^4 q / t^4 E)$. This parameter may be considered as one measure of the significance of the large-deflection behaviour undergone by a plate, where in most cases a value of Q as low as 5 warrants a large-deflection design.

(2) Boundary conditions

Under the same loading and geometry, plates with different boundary conditions show different degrees of large-deflection behaviour. Plates restrained against in-plane movement at the supports generally show higher large-deflection effects.

(3) Material properties

Yielding of plate, or initiation of fracture, are the basis of limit state design criteria for ductile and brittle materials respectively. The higher the yield strength or the fracture stress, the more necessary it is to design the plate for large deflections.

1.2 Large-deflection behaviour

For a reader not familiar with plate theory, the large-deflection of plates and its associated phenomena can be best illustrated by making use of simple beam examples. Consider the simple plane grid shown in Fig. 1.1(i), consisting of six horizontal beams pinned together at the corners A, B, C and D. The grid is supported on four roller supports at the four corners, and is loaded at its centre point E by a concentrated load P. It is clear that, due to symmetry, the reaction at each support is an upward force equal to $P/4$.

For small values of P the free-body diagram of a diagonal and a side member is shown in Fig. 1.1(iii). The diagonal members carry the entire loading to the supports by their bending action. The side members make no load-carrying contribution. The relationship between the load P and the out-of-plane deflection of the diagonals is linear. This system of load distribution and the resulting behaviour is accurate for low values of deflection and is called small-deflection behaviour. The behaviour is shown by a straight line in Fig. 1.2. The key assumption in this case is that the equations of equilibrium and compatibility are written for the undeformed shape of the grid. The change of geometry due to deformation is considered to be too small to affect these equations significantly.

As P and consequently deflections increase, the out-of-plane deflection w of the diagonals tends to pull the corners of the grid towards the centre in the horizontal plane. If the side members were not present, the shortening of distance between the two ends of each diagonal could be easily accommodated by the movement of the rollers, with practically no change in the linear relationship between the transverse load P and out-of-plane deflection w. However, the side members in this example act as a restraint against the movement of the corners in the plane of the grid, as a result of which they develop axial compressive forces coupled with tensile equilibrating axial forces in the diagonals. The axial force diagrams of the diagonal and side members are shown in Fig. 1.1(iv). Figure 1.1(v) illustrates the general pattern of the axial forces developed in the grid members. Clearly, at this stage, the relationship between the value of transverse load P and the central out-of-plane deflection w is no longer linear. In general, there is a reduction in the rate of increase in deflection and the related bending stresses depending on the type of the structure. This non-linear behaviour is shown in Fig. 1.2 by the curves designated large deflection. Note that the non-linearity is due to change of geometry

4 Thin Plate Design For Transverse Loading

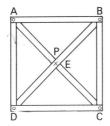
(i) Plan view of the six-member plane grid

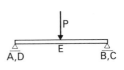
(ii) Side elevation of the plane grid

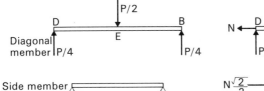
(iii) Small-deflection free body diagram

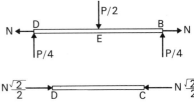
(iv) Large-deflection free body diagram

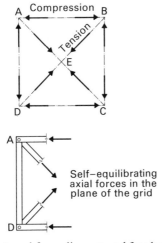
(v) Internal-force diagram and free body diagram of a section of grid

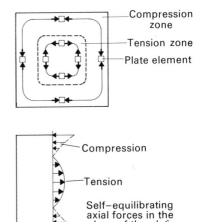
(vi) Internal membrane force diagram and section of a simply supported square plate

Fig. 1.1. Large-deflection axial forces in a plane grid and a square plate.

General Considerations 5

of the grid under loading. At each value of loading, the related deflections and the stresses may be calculated by allowing for the change of geometry in the equations.

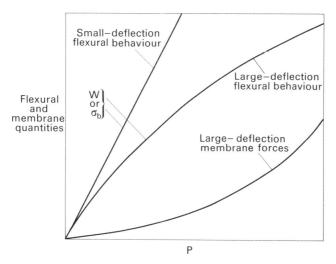

Fig. 1.2. Characteristic variations of small-deflection and large-deflection quantities with the transverse loading.

Now, a square plate under a central concentrated loading P and supported on rotationally free roller supports acts in basically the same manner (simply supported) as the plane grid. As the loading is increased the plate passes through the purely flexural small-deflection range to the non-linear large-deflection range. At higher loads, similarly to the plane grid, a central tensile zone develops, within which the plate elements are all under axial tension, as well as bending and shearing stresses (see Fig. 1.1(vi)). The central tensile zone is in equilibrium with an outer compression ring. The forces developed in the plane of the plate correspond to the axial forces in the members of the grid and are called membrane forces. Figure 1.2 shows typical variations of flexural and membrane forces with the transverse load P.

1.3 Historical background

Although the equations of the large deflection of plates were derived in 1910 (Von Karman 1910), it is only through recent advances in the development of numerical methods that the general problem of plates has been treated satisfactorily. The early investigators used the Ritz energy method (Way 1938), infinite double Fourier series (Levy 1942) and finite differences (Kaiser 1936, Way 1938) to obtain solutions for a number of isolated cases of plates under uniformly distributed loading and simple boundary conditions.

Using finite differences and modern computational techniques, Basu and Chapman (1966) restudied the problem and obtained several solutions for new boundary conditions. Developing a refined finite-difference formulation and solution procedure, Aalami and Chapman (1969a) made a comprehensive study of the large-deflection behaviour of plates under transverse and in-plane loading.

The emergence of the dynamic relaxation method (Otter 1965), and its development into a general numerical technique for the solution of partial differential equations (Otter and Cassell 1966), led to further solutions for particular large-deflection plate problems (Rushton 1969a,b, Williams 1971), in particular for initially deformed plates subject to in-plane loading.

As part of the general development of finite-element techniques, a number of diversified treatments have been proposed (Mallet and Marcal 1968, Murray and Wilson 1969, Zienkiewicz 1971) for the large-deflection behaviour of plates, by means of which a number of typical plate problems have been successfully analysed.

There are a number of recent treatments of the subject in regard to orthotropic plates, different boundary conditions, different loading conditions and shapes of plates as reported by Berger (1955), Weiss (1969), and Chia (1972).

Work is also in progress in the field of large-deflection elastoplastic analysis of plates, in which yielding of the material and the development of plastic zones are treated (Jaeger 1958, Wah 1958, Massonnet 1968, Lin *et al.* 1972, Crisfield 1973).

1.4 Basic plate relationships and equations

1.4.1 Sectional actions

Consider a homogeneous rectangular isotropic plate with sides a and b and thickness t as shown in Fig. 1.3. At any point at distances x and y from the coordinate axes, the acting components of forces and

General Considerations 7

moments may be resolved into bending moments M_x and M_y, twisting moments M_{xy}, normal shearing forces Q_x and Q_y, direct membrane forces N_x and N_y and finally membrane shearing forces N_{xy}. These components are collectively called *sectional actions* and are normally grouped into those producing out-of-plane deformations (M_x, M_y, M_{xy}, Q_x, Q_y) and those giving rise to in-plane deformations of the mid-plane of the plate (N_x, N_y, N_{xy}).

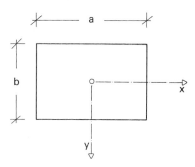

Fig. 1.3. Coordinate axes.

Components of the first group are referred to as *flexural actions*, and together with the related transverse pressure are shown in Fig. 1.4(i) acting on an element of plate with their positive directions. The latter group, called *membrane actions*, are shown in Fig. 1.4(ii) also with their positive directions. Of course, these components all act on the same element, but for clarity are demonstrated separately. Note that in the figures double arrows are used as vectorial representation of moments in the direction of the arrow and in the sense given by the right-hand screw rule. For example, in Fig. 1.4(i) M_x and M_y are moments causing tension at the bottom face of the element.

1.4.2 Stresses on an element of plate

Each of the sectional actions introduced corresponds to a distribution of reacting stress through the depth of the plate. The relation between each action and the corresponding distribution of stress on the element face is such that the summation of the stresses on that face equals the action concerned. The distribution of stresses on an element of plate due to the positive direction of the sectional actions is shown

8 Thin Plate Design For Transverse Loading

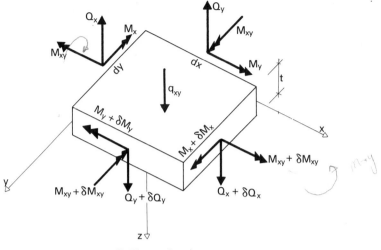

(i) Flexural actions

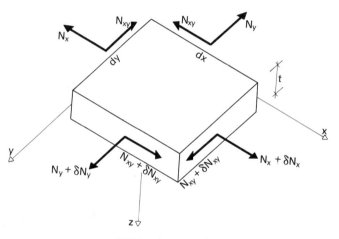

(ii) Membrane actions

Fig. 1.4. Sectional actions.

in Fig. 1.5. For any given value of loading, the complete state of stress of the plate element is the algebraic sum of the stresses shown. For example, the total direct stress at the lowest fibre of the element and on the face perpendicular to the x-axis and distance dx from it, can be seen from Fig. 1.5 to be equal to $\sigma_{bx} + \sigma_{mx}$.

General Considerations 9

Among the stress components shown in Fig. 1.5, stresses τ_{xz}, and τ_{yz} due to Q_x and Q_y respectively are in most practical cases insignificant compared to the other components of stress, and are normally disregarded in design.

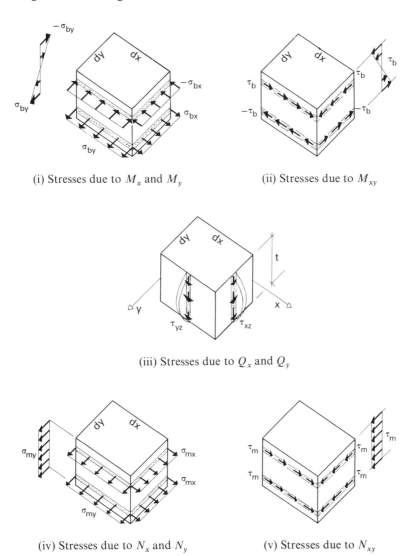

(i) Stresses due to M_x and M_y

(ii) Stresses due to M_{xy}

(iii) Stresses due to Q_x and Q_y

(iv) Stresses due to N_x and N_y

(v) Stresses due to N_{xy}

Fig. 1.5. Stresses on an element of plate.

10 Thin Plate Design For Transverse Loading

(1) Relationship between sectional actions and the related stresses

Maximum bending stresses at lowermost fibre due to M_x and M_y

$$\sigma_{bx} = 6M_x/t^2$$
$$\sigma_{by} = 6M_y/t^2 \qquad (1.1)$$

Maximum membrane shearing stress due to M_{xy}

$$\tau_b = -6M_{xy}/t^2 \qquad (1.2)$$

Maximum shearing stresses due to Q_x and Q_y

$$\tau_{xz} = 1\cdot5 Q_x/t$$
$$\tau_{yz} = 1\cdot5 Q_y/t \qquad (1.3)$$

Average direct membrane stresses and membrane shearing stresses due to N_x, N_y and N_{xy}

$$\left.\begin{array}{l}\sigma_{mx} = N_x/t \\ \sigma_{my} = N_y/t \\ \tau_m = N_{xy}/t\end{array}\right\} \qquad (1.4)$$

henceforth, stresses τ_{xz} and τ_{yz} will be disregarded, because they are not critical to the cases considered in this book.

(2) Total stresses

Maximum total stresses (σ_x, σ_y, τ) in the x- and y-directions occur either at the lowermost or the uppermost fibre of the element. From Fig. 1.5 these are as follows:

at the lowermost fibre

$$\left.\begin{array}{l}\sigma_x = \sigma_{bx} + \sigma_{mx} \\ \sigma_y = \sigma_{by} + \sigma_{my} \\ \tau = -\tau_b + \tau_m\end{array}\right\} \qquad (1.5)$$

at the uppermost fibre

$$\left.\begin{array}{l}\sigma_x = -\sigma_{bx} + \sigma_{mx} \\ \sigma_y = -\sigma_{by} + \sigma_{my} \\ \tau = \tau_b + \tau_m\end{array}\right\} \qquad (1.6)$$

1.4.3 Basic plate relationships

For a homogeneous isotropic plate, the basic relationships for stresses and deformations of a plate element as shown in Fig. 1.6 may be summarised as follows.

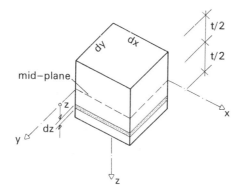

Fig. 1.6. An element of plate.

(1) Small deflections—flexural behaviour

The behaviour of plates in which the mid-plane undergoes only out-of-plane deformation (w in the z-direction) is governed by the so-called small-deflection relationships and equations. For these plates the relationships of stresses and deformations with the transverse loading are linear.

The stresses and deformations of an element of plate due to the flexural actions alone (M_x, M_y, M_{xy}, Q_x, Q_y) are related to the transverse out-of-plane deflection of plate w by the following relationships.
Rotation of mid-plane

$$\text{In } x\text{-direction} \quad w_{,x}$$
$$\text{In } y\text{-direction} \quad w_{,y} \tag{1.7}$$

(Note that a comma followed by subscripts represents partial differentiation in turn with respect to each subscript variable.)

Displacement of a point on a layer distance z from the mid-plane (Fig. 1.6)

$$u = -zw_{,x}$$
$$v = -zw_{,y} \tag{1.8}$$

12 Thin Plate Design For Transverse Loading

Hence strains at the same layer

$$\left.\begin{aligned}\varepsilon_x &= u_{,x} = -zw_{,xx}\\ \varepsilon_y &= v_{,y} = -zw_{,yy}\\ \gamma_{xy} &= u_{,y} + v_{,x} = -2zw_{,xy}\end{aligned}\right\} \quad (1.9)$$

Bending stresses at the same layer

$$\left.\begin{aligned}\sigma_x &= \frac{E}{(1-v^2)}(\varepsilon_x + v\varepsilon_y) = \frac{-zE}{(1-v^2)}(w_{,xx} + vw_{,yy})\\ \sigma_y &= \frac{E}{(1-v^2)}(\varepsilon_y + v\varepsilon_x) = \frac{-zE}{(1-v^2)}(w_{,yy} + vw_{,xx})\\ \tau_b &= G\gamma_{xy} = \frac{-zE}{(1+v)}w_{,xy}\end{aligned}\right\} \quad (1.10)$$

From integration of the stresses on the faces of the element and the equilibrium considerations (Timoshenko and Woinowsky-Krieger 1959) the flexural actions are

$$\left.\begin{aligned}M_x &= -D(w_{,xx} + vw_{,yy})\\ M_y &= -D(w_{,yy} + vw_{,xx})\\ M_{xy} &= D(1-v)w_{,xy}\\ Q_x &= -D(w_{,xxx} + w_{,xyy})\\ Q_y &= -D(w_{,yyy} + w_{,xxy})\end{aligned}\right\} \quad (1.11)$$

(2) Membrane behaviour

For purely membrane behaviour, the plate does not deform out of its plane. The deformations are confined to the x- and y-directions only, and are constant through the thickness of the plate for all the layers. There are several ways of expressing the membrane forces and stresses due to the membrane actions N_x, N_y and N_{xy}. Herein Airy's stress function, which is defined below, is employed for the expression of the membrane actions, stresses and deformations.

Airy's stress function f is defined as

$$\left. \begin{array}{l} N_x = t\sigma_{mx} = tf_{,yy} \\ N_y = t\sigma_{my} = tf_{,xx} \\ N_{xy} = t\tau_m = -tf_{,xy} \end{array} \right\} \quad (1.12)$$

Mid-plane strains in terms of forces and stress function

$$\left. \begin{array}{l} \varepsilon_x = u_{,x} = \dfrac{1}{Et}(N_x - vN_y) = \dfrac{1}{E}(f_{,yy} - vf_{,xx}) \\ \varepsilon_y = v_{,y} = \dfrac{1}{Et}(N_y - vN_x) = \dfrac{1}{E}(f_{,xx} - vf_{,yy}) \\ \gamma_{xy} = u_{,y} + v_{,x} = \dfrac{2(1+v)}{Et}N_{xy} = -\dfrac{2(1+v)}{E}f_{,xy} \end{array} \right\} \quad (1.13)$$

(3) Large-deflection behaviour—interaction between the flexural and membrane actions

In large-deflection behaviour the interaction between flexural and membrane actions is taken into account. In this case deflections and stresses vary in a non-linear manner with the magnitude of the transverse pressure.

For large deflections with out-of-plane deformations up to several times the plate thickness, the mid-plane strains should be modified in the following form (Von Karman 1910).

$$\left. \begin{array}{l} \varepsilon_x = u_{,x} + 0\cdot5(w_{,x})^2 = \dfrac{1}{E}(f_{,yy} - vf_{,xx}) \\ \varepsilon_y = v_{,y} + 0\cdot5(w_{,y})^2 = \dfrac{1}{E}(f_{,xx} - vf_{,yy}) \\ \gamma_{xy} = u_{,y} + v_{,x} + w_{,x}w_{,y} = -\dfrac{2(1+v)}{E}f_{,xy} \end{array} \right\} \quad (1.14)$$

The expressions given for the sectional actions in terms of out-of-plane deformation w and the stress function f are equally valid in the large-deflection range. In other words, bending moments and normal

14 Thin Plate Design For Transverse Loading

shearing forces are expressed by the relationship (1.11) and the membrane actions are given by (1.12). In the relationships discussed herein, the effects of shearing stresses τ_{xz} and τ_{yz} due to Q_x and Q_y are neglected due to their insignificant influence in homogeneous isotropic plates considered in this book.

1.4.4 Governing plate equations

For a flat isotropic plate, the derivation of the governing equations is described in detail in Timoshenko and Woinowsky-Krieger (1959). In the following, the equations as used in this book are quoted.
For small deflections

$$w_{,xxxx} + 2w_{,xxyy} + w_{,yyyy} = q_{xy}/D \tag{1.15}$$

For membrane action alone

$$f_{,xxxx} + 2f_{,xxyy} + f_{,yyyy} = 0 \tag{1.16}$$

For large deflections, the behaviour is expressed by two fourth-order partial differential equations in terms of mid-plane out-of-plane deflection w and Airy's stress function f.

$$w_{,xxxx} + 2w_{,xxyy} + w_{,yyyy} - \frac{t}{D}(f_{,yy}w_{,xx} - 2f_{,xy}w_{,xy} + f_{,xx}w_{,yy})$$

$$= q_{xy}/D \tag{1.17}$$

$$f_{,xxxx} + 2f_{,xxyy} + f_{,yyyy} = E[(w_{,xy})^2 - w_{,xx}w_{,yy}] \tag{1.18}$$

1.4.5 Boundary conditions

(1) Flexural boundary conditions

For rotationally free edges
At $x = \pm a/2$, $M_x = 0$, which gives

$$w_{,xx} + vw_{,yy} = 0 \tag{1.19}$$

At $y = \pm b/2$, $M_y = 0$, which gives

$$w_{,yy} + vw_{,xx} = 0 \tag{1.20}$$

For rotationally fixed edges, rotation across the boundary is equated to zero

At $x = \pm a/2$

$$w_{,x} = 0 \tag{1.21}$$

At $y = \pm b/2$

$$w_{,y} = 0 \tag{1.22}$$

For edges on rigid supports, both at $x = \pm a/2$ and $y = \pm b/2$

$$w = 0 \tag{1.23}$$

For edges with no normal support (no support in z-direction)

At $x = \pm a/2$

$$w_{,xxx} + (2 - v)w_{,xyy} = 0 \tag{1.24}$$

At $y = \pm b/2$

$$w_{,yyy} + (2 - v)w_{,xxy} = 0 \tag{1.25}$$

(2) Membrane boundary conditions

For defining the edge displacement of the boundaries perpendicular to the line of support and in the plane of the plate any one of the following three alternatives may be applicable.

(a) For stress-free edges

At $x = \pm a/2$, $\sigma_{mx} = 0$, which gives

$$f_{,yy} = 0 \tag{1.26}$$

At $y = \pm b/2$, $\sigma_{my} = 0$, which gives

$$f_{,xx} = 0 \tag{1.27}$$

(b) For zero displacement of the edge across the boundary (edge remains straight and in position)

At $x = \pm a/2$

$$u = \int u_{,x} \, dx = \frac{1}{E} \int \left[(f_{,yy} - vf_{,xx}) - 0.5(w_{,x})^2 \right] dx = 0$$

$$\tag{1.28}$$

16 Thin Plate Design For Transverse Loading

At $y = \pm b/2$

$$v = \int v_{,y}\, dy = \frac{1}{E}\int [(f_{,xx} - vf_{,yy}) - 0{\cdot}5(w_{,y})^2]\, dy = 0 \tag{1.29}$$

(c) For the edge remaining straight and displacing perpendicular to the line of support in the plane of the plate, with the integral of the membrane direct stress along the edge equal to zero

At $x = \pm a/2$

$$u = \frac{1}{E}\int [(f_{,yy} - vf_{,xx}) - 0{\cdot}5(w_{,x})^2]\, dx = \text{constant} \tag{1.30}$$

$$\int_{-b/2}^{+b/2} t\sigma_x\, dy = t\int_{-b/2}^{+b/2} f_{,yy}\, dy = 0 \tag{1.31}$$

At $y = \pm b/2$

$$v = \frac{1}{E}\int [(f_{,xx} - vf_{,yy}) - 0{\cdot}5(w_{,y})^2]\, dy = \text{constant} \tag{1.32}$$

$$\int_{-a/2}^{+a/2} t\sigma_y\, dx = t\int_{-a/2}^{+a/2} f_{,xx}\, dx = 0 \tag{1.33}$$

For the membrane shear boundary conditions the following two cases are considered. For the stress-free edges, both at $x = \pm a/2$ and $y = \pm b/2$ the membrane shearing stress τ_m is equated to zero, hence

$$-f_{,xy} = 0 \tag{1.34}$$

For the extreme case of the plate being connected to a rigid boundary such that no displacement of the boundary along the line of support may take place

At $x = \pm a/2$, $v_{,y} = 0$, which gives

$$f_{,xx} - vf_{,yy} = 0{\cdot}5E(w_{,y})^2 \tag{1.35}$$

Similarly at $y = \pm b/2$, $u_{,x} = 0$

$$f_{,yy} - vf_{,xx} = 0{\cdot}5E(w_{,x})^2 \tag{1.36}$$

1.4.6 Outline of the solution procedure employed

The governing equations and the boundary conditions are expressed in terms of finite-difference approximations, and the resulting sets of non-linear algebraic equations are solved by an iterative method and a digital computer as outlined by Aalami (1969a). The approximations are derived in a general form for a graded mesh, which increases the accuracy of the results (for a given number of nodal points) when the plate is subjected to concentrated loads, or when the edges are rotationally fixed.

In obtaining the results, the complete plate is represented by 300–400 nodes for which an accuracy condition is specified in the iteration, such that numerical values are within 0·5% of the values to which the solution would converge. Comparisons with existing large-deflection solutions indicate that the accuracy of the present solutions is good (Aalami 1969b).

CHAPTER TWO
Large-deflection Design

2.1 Large-deflection design considerations for plates under transverse loading

A large-deflection design or check for a panel of plating under transverse loading (loading perpendicular to the plane of the plate) consists of the following steps:

(1) Definition of the panel geometry, boundary conditions and loading.
(2) Selection of design criteria and material.
(3) Evaluation of plate thickness and/or other selected design parameters.

2.2 Geometry, boundary conditions and loading

2.2.1 Geometry

The plates considered herein are of rectangular or circular geometry. For the rectangular range of plates, aspect ratios (ratio of length b to width a) considered cover in most cases the range of 0·67 to 3 and the intermediate ratios of 1, 1·5 and 2. The results of small-deflection plate analyses indicate that for plates acted upon by a uniform distribution of pressure or a central concentrated loading, it is sufficient to treat aspect ratios up to 2–3 for design data. For the purpose of design, maximum deflections and stresses of longer plates having aspect ratios greater than the quoted range are found to be reasonably close to the plates with aspect ratios 2–3, depending on the loading and boundary conditions.

For the large-deflection design, on the basis of conclusions drawn from small-deflection analyses, it is proposed to design the panels with aspect ratios greater than 2–3 as plates having aspect ratios 2–3 with the same boundary conditions, but naturally acted upon only by that proportion of loading which falls within the boundaries of the smaller equivalent plate considered. It is assumed that the large-deflection quantities thus obtained are a close approximation of the related values of the actual aspect ratio, and may be accepted for design.

For plates with aspect ratios between the values treated, it is proposed to evaluate the design quantities for two plates, one having a higher and the other a lower aspect ratio, but both with the same boundary conditions. The design quantities for the plate under investigation are to be linearly interpolated between the two values calculated. For example, the deflection of a plate with aspect ratio 1·85 (not treated herein) is estimated from linear interpolation of maximum deflections of plates with aspect ratios 1·5 and 2 (both treated herein) under the same loading. Obviously the variations of deflections and sectional actions with aspect ratio are not linear, but the suggested linear interpolation is considered to be a simple and yet accurate enough approximation to be used in design.

2.2.2 Boundary conditions

(1) General

The overall design dimensions and aspect ratio having been determined, the design boundary conditions of the plate will now be clearly defined. It should be noted that plate panels do not act in isolation, but often, as part of a more complex assembly, they act together with the adjoining structural elements in sustaining the applied loading. Since the integrated design of a structural assembly such as a stiffened panel, consisting of the skin, stiffeners and the web, is complicated and time-consuming, it is not going to be the everyday practice of the majority of design offices—at least not in the near future. A simple and rational approach to this complicated problem is to consider the plate panels in isolation for design. Provided the interaction of the isolated panel with the adjoining structure is well understood and appropriate boundary conditions are specified, the design will be realistic.

For the large-deflection design, four boundary conditions need to be specified at each of the plate edges, two relating to its flexural behaviour and two to its membrane (in-plane) conditions. Consider a flexural boundary condition such as the rotation across a boundary. There are two extreme limits for this boundary condition. One is the condition of the edges rotationally fixed (clamped case), and the other is the case of the edges being rotationally free (simple support). In dealing with an actual design problem, often the relating boundary conditions are neither of the two extremes mentioned. In such cases the designer should either decide on the extreme condition which results in the more critical design quantities, or should consider both extremes separately and proportion the results on the judgement of the actual conditions and their relation to each of the two extremes. Numerical examples in Chapter 4 illustrate this point.

(2) Boundary conditions considered

In this book, of the extreme flexural and membrane boundary conditions possible, only the cases which are considered to be of more practical significance are treated. For ease of definition, quick recognition and speedy reference, the boundary conditions considered are expressed symbolically on design curves and tables given. Symbols used in the representation of the boundary conditions are listed in Fig. 2.1. The mathematical formulations for each of the cases shown in this figure have already been discussed in Chapter 1.

For each plate problem, the boundary conditions are usually represented by two symbolic figures (rectangles or circles). The first figure, which is located on the left-hand side and normally bears the Cartesian coordinates, represents the plate's flexural boundary conditions (FBC), and the second figure shows the related membrane boundary conditions (MBC).

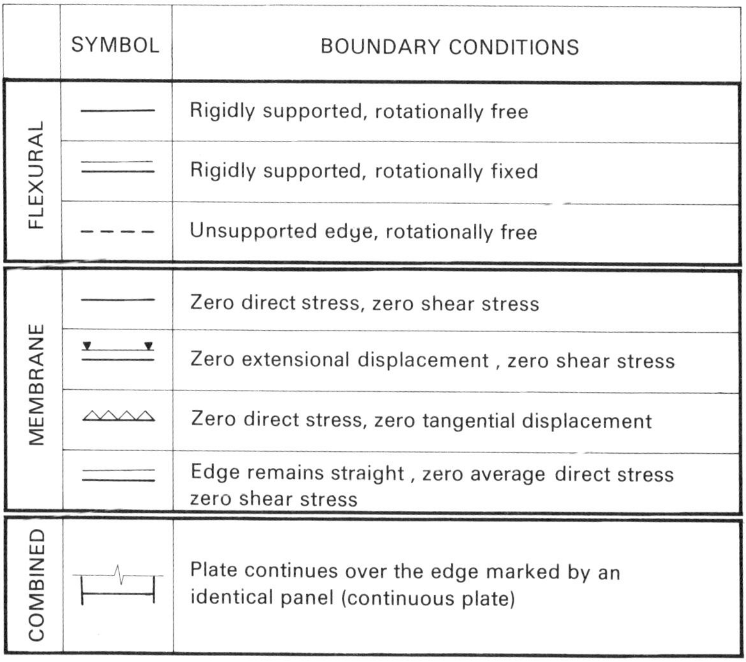

	SYMBOL	BOUNDARY CONDITIONS
FLEXURAL	————	Rigidly supported, rotationally free
	════	Rigidly supported, rotationally fixed
	----	Unsupported edge, rotationally free
MEMBRANE	————	Zero direct stress, zero shear stress
	▼——▼	Zero extensional displacement, zero shear stress
	△△△△	Zero direct stress, zero tangential displacement
	════	Edge remains straight, zero average direct stress, zero shear stress
COMBINED	⊢∧⊣	Plate continues over the edge marked by an identical panel (continuous plate)

Fig. 2.1. Legend of basic symbols used for representation of flexural and membrane boundary conditions.

(3) Examples for symbolic representation of boundary conditions

In order to illustrate the use of the symbols of Fig. 2.1 in describing the boundary conditions of a plate, consider the examples shown in Fig. 2.2.

For Fig. 2.2(i):

Flexural: at $x = \pm a/2$ and $y = \pm b/2$, edges are rigidly supported and rotationally free.

Membrane: at $x = \pm a/2$ and $y = \pm b/2$, zero direct membrane stress and zero shear stress.

This plate corresponds to a truly isolated single panel on rigid roller supports.

For Fig. 2.2(ii):

Flexural: at $x = \pm a/2$, edges are rigidly supported and are rotationally fixed. At $y = \pm b/2$, edges are rigidly supported and are rotationally free.

Membrane: at $x = \pm a/2$, zero direct membrane stress and zero membrane shear stress (edges free for in-plane movement). At $y = b/2$, zero extensional displacement in the y-direction, together with zero membrane shear stress. At $y = -b/2$, zero direct membrane stress, together with zero tangential displacement in the x-direction, i.e., the points on this support are free to move only in the y-direction

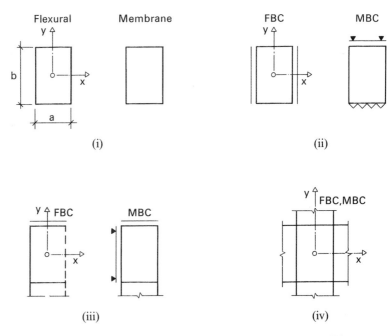

Fig. 2.2. Examples of symbolic representation of boundary conditions.

For Fig. 2.2(iii):
Flexural: at $x = a/2$, the edge is unsupported and is rotationally free.
At $x = -a/2$, the edge is rigidly supported and is rotationally free.
At $y = b/2$, the edge is rigidly supported and is rotationally fixed.
At $y = -b/2$, the plate panel is extended beyond this support by a loaded identical panel.
Membrane: at $x = a/2$, zero direct membrane stress and zero membrane shear stress.

At $x = -a/2$, zero extensional displacement in the x-direction, together with zero membrane shear stress.

At $y = b/2$, the edge remains straight and displaces parallel to the x-axis with the average membrane direct stress on it equal to zero, together with zero membrane shear stress.

At $y = -b/2$, the plate panel is continuous over this support as a loaded identical panel.

For Fig. 2.2(iv):

Flexural and membrane: The plate panel is continuous over all four edges by identical panels forming a total of 3×3 (9) panels, from which, in the case of concentrated loads, only the central panel (shown in the figure) is loaded. In the case of uniformly distributed loading, all the panels are loaded.

(4) Examples of boundary-conditions idealisations

The following examples illustrate some of the possible ways of idealising the boundary conditions of actual plate panels under transverse loading. It should be noted that for plates under in-plane action, or the combined actions of in-plane and transverse loadings, idealisations other than those given in the examples may be satisfactory.

Example 2.1

One of the simplest idealisations is the case of a temporary steel sheet covering a service opening or a trench on a road, while allowing for an undisturbed flow of traffic. As is shown in Fig. 2.3, the idealised

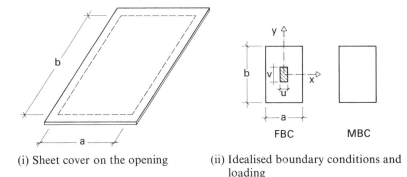

(i) Sheet cover on the opening

(ii) Idealised boundary conditions and loading

Fig. 2.3. Temporary sheet cover of a service opening on a road.

24 Thin Plate Design For Transverse Loading

plate may be assumed to have the dimensions of the opening and be freely resting on rigid supports with the in-plane membrane direct and shearing stresses equal to zero. For such plates the critical design loading is usually a wheel load assumed as a uniform pressure over a patch area. Fig. 2.3(ii) shows symbolically the flexural boundary conditions (FBC) and the membrane boundary conditions (MBC) by means of two rectangles as described in (2) of this section.

Example 2.2
Panes of glass and other thin covering plates may be idealised under wind pressure or sonic boom as simply supported plates with no in-plane restraints, as shown in Fig. 2.4(ii).

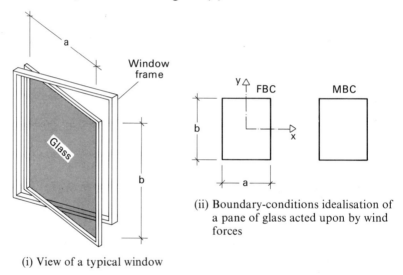

(i) View of a typical window

(ii) Boundary-conditions idealisation of a pane of glass acted upon by wind forces

Fig. 2.4. Boundary-conditions idealisations of window glass panels.

It is frequently the case that the standard fixtures used for installation of panes of glass and other covers exert some degree of rotational restraint to the panels which is not accounted for in the idealisation. Unless the fixtures achieve near rotational fixity at the boundaries, the deflections and stresses given by the proposed idealisation are on the safe side.

Example 2.3
For the plate cover of a submerged inspection door on a barrier, dock

gate or similar structure with a construction as shown in Fig. 2.5, the rotational restraint of the edge beam may be disregarded, since the practical details and sizes are such that appreciable fixed end moments do not develop. For the membrane boundary conditions, each of the edges is restrained by the edge beam provided at that edge against movements both perpendicular and parallel to the line of support.

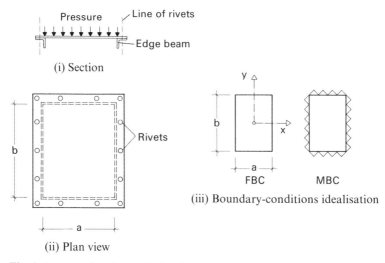

Fig. 2.5. Inspection door of a barrier.

The edge beams bend in the plane of the plate between the corners of the door and undergo compression along their lengths. Practical dimensions are normally such that the resistance to the movement of the plate along the supports is stronger than the resistance to the movement perpendicular to them. As an upper limit, each boundary is idealised to be fully restrained against in-plane tangential movement parallel to itself (along the boundary) with zero direct stress, as is shown diagrammatically in Fig. 2.5(ii).

Example 2.4
The side walls of cubic or long rectangular liquid containers can be idealised in several ways, depending on the constructional details. Figure 2.6 shows several possible alternatives.

The rotational fixity of the side edges of the tank walls is due to symmetry. The rotational fixity idealisation of the bottom edge of a side wall is not so clear. It depends on the ratio of the plate thicknesses of the side walls and the bottom. It also depends on the way in which the tank is supported at the bottom. In many cases the liquid pressure on the tank bottom results in a near-fixity rotational condition at the bottom edge.

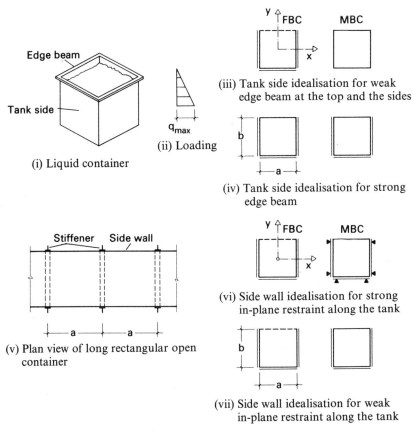

Fig. 2.6. Some boundary-condition idealisations of liquid containers.

Note that the in-plane tensile forces on the edges of the tank walls, which react with the pressure on the adjacent sides, are not directly accounted for. In the boundary-condition idealisations, it is assumed that these forces are taken wholly by the corner stiffeners and the supporting frame.

For the required stiffness of the edge beam at the top of the side wall, refer to Section 4.4.

If the side edges of a tank wall are not provided with stiffeners (as is the case in Fig. 2.6(i)), the membrane boundary conditions for these edges ($x = \pm a/2$) lie between the limiting cases of zero direct membrane stress (Fig. 2.6(iii)), and edges remaining straight (Fig. 2.6(iv)), possibly closer to the former condition.

Example 2.5
In the floating platform shown in Fig. 2.7, provided the distribution of loading is fairly uniform there is no overall bending of the platform. The panels of the bottom shell spanning between the webs (Fig. 2.7(ii)) are acted upon by a uniform distribution of water pressure.

Depending on the position of the panel on the bottom shell, the suggested boundary-condition idealisations are shown in Figs. 2.7(iii)–(vi). There are two possible limiting boundary-condition idealisations of panels C. In Fig. 2.7(v), the in-plane restraint provided by the webs is neglected and that of the adjacent panels allowed for. In Fig. 2.7(vi) only the in-plane restraint of the webs is considered in a limiting form.

Example 2.6
Figure 2.8 shows the view of a section of a dry cargo carrier. Apart from panel A, for which wheel loading from a fork-lift truck may prove more critical than the in-service loading of the cruising carrier, for the rest of the panels the relating in-plane loadings caused by the overall bending of the ship and/or the local bending of the bottom between the sides and the bulkhead should also be accounted for. The boundary-condition idealisations shown apply to typical panels positioned so as to be subjected to insignificant in-plane loading. Transverse loading, and hence the proposed idealisation for panel D, is primarily for tankers and for cases in which alternate compartments are loaded. In the general case, the loading and the boundary conditions of most ship plate panels are governed by the presence of in-plane loading.

Example 2.7
Figure 2.9 shows the possible boundary-condition idealisations of an orthotropic steel bridge deck acted upon by a wheel load. The

28 Thin Plate Design For Transverse Loading

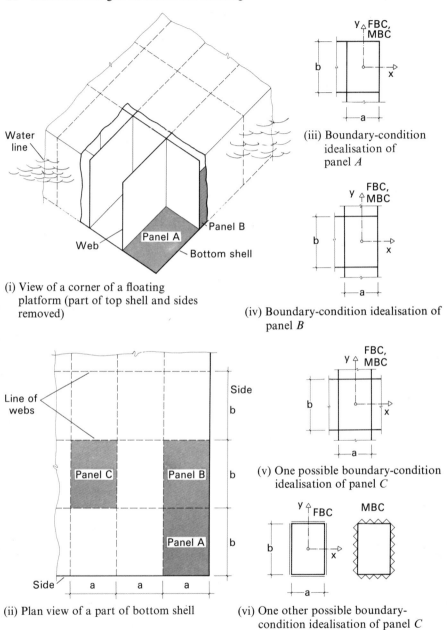

(i) View of a corner of a floating platform (part of top shell and sides removed)

(ii) Plan view of a part of bottom shell

(iii) Boundary-condition idealisation of panel A

(iv) Boundary-condition idealisation of panel B

(v) One possible boundary-condition idealisation of panel C

(vi) One other possible boundary-condition idealisation of panel C

Fig. 2.7. Boundary-condition idealisations of bottom shell panels of a floating platform.

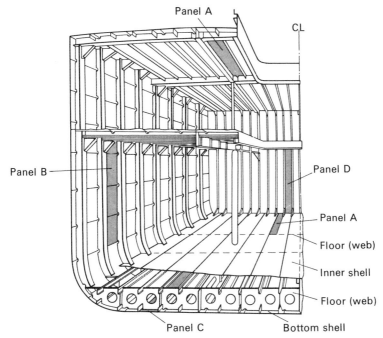

(i) Sectional view of a dry cargo carrier showing some of the typical panels of plating [from Barabanov (undated)]

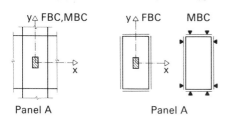

(ii) Possible boundary-condition idealisations of panel A under wheel loading from a fork-lift truck

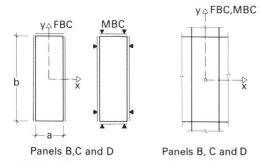

(iii) Possible idealisations of panels B, C and D under transverse pressure alone. Panel D is normally acted upon by a transverse pressure in tankers only. When considering the simultaneous actions of transverse pressure and in-plane loading, other boundary-condition idealisations may become necessary

Fig. 2.8. Boundary-condition idealisations of some ship plate panels.

30 Thin Plate Design For Transverse Loading

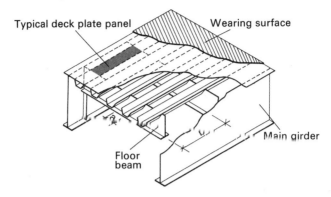

(i) Orthotropic steel bridge deck with torsionally stiff ribs

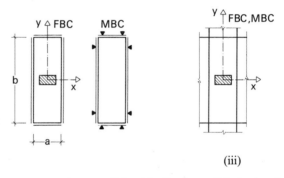

(iii)

(ii) and (iii) Possible boundary-condition idealisations of deck plate between the rib webs for the evaluation of local deck-plate stresses under wheel loading

Fig. 2.9. Steel bridge decks. Panel boundary-condition idealisation for local stresses.

proposed idealisations may be used for the evaluation of local stresses in the deck plate.

Example 2.8
Plate panels on the tip of the aircraft wings are predominantly under transverse pressure. The boundary-condition idealisations of the individual panels may be based on either (i) symmetry considerations of the panels and hence the postulation that the edges remain in position, or (ii) the resistance of the webs and the stiffeners to in-plane shearing movement of the edges along the boundaries. The idealisa-

tions (i) and (ii) are shown in Fig. 2.10. Alternatively the panels may be assumed to be continuous on four sides over the supports (idealisation not shown in the figure).

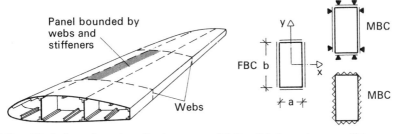

(i) Simplified view of an aircraft wing

(ii) Possible boundary-condition idealisations for transverse loading

Fig. 2.10. Boundary-condition idealisations for aircraft panels under transverse pressure.

2.3 Selection of design criteria

The governing design criteria in each case should be selected such that with an appropriate factor of safety, the plate would reach its limit of unserviceability under the factored in-service loading. The selection of a limit of unserviceability is based on the material properties of the plate and the plate function. As an example, for plating used in bridge construction, Merrison *et al.* (1973) have laid out the related design criteria. Similarly for glass panels a set of design criteria are suggested by PPG Industries (undated).

For the general case, the design criteria may be categorised as follows:

(1) Control of deformations.
(2) Control of first surface yield, or initiation of cracking.
(3) Allowance for limited spread of yield through the depth under working conditions.
(4) Allowance for localised penetration of yield to the mid-plane of the plate (formation of plastic hinges).

2.3.1 Control of deformations

In thin-plate design, maximum allowable deflection is normally specified as a ratio of plate thickness. Depending on the plate function, the allowable out-of-plane deflection varies between a fraction of the thickness to several times the thickness. However, where the plate is furnished with a wearing surface, such as a layer of asphalt on an orthotropic steel bridge deck, it is customary to specify the deflection as a ratio of the plate side (span between the ribs) to avoid the cracking of the wearing surface and the subsequent rusting of the steel plate. Naturally, for control of excessive undulations of a surface under wheel loading, the deflection should again be controlled with respect to the plate sides. In the preceding examples, the permissible out-of-plane deflection varies between 1/200 to 1/400 of the spacing between the stiffeners.

The protective coatings of some liquid chemical containers start cracking at given limiting magnitudes of tensile strains. In such cases the maximum tensile strain is to be specified and controlled for the coated plate components forming these containers.

For panes of glass (example of Fig. 2.4) out-of-plane deflections several times the plate thickness may occur under working conditions. Liptak et al. (1973) quote deflection values well over five times the thickness for larger plate panels. Where glass is used for overhead lighting or on a slope, it may be necessary to check out-of-plane deformations against ponding of water.

For a liquid container, such as the example of Fig. 2.6, made of metal sheets and used as a chemical bath in industry, deflections several times the plate thickness may be tolerated under working conditions.

For the bottom panels of the floating platform of Fig. 2.7, where the plates are under transverse loading alone, the magnitude of the out-of-plane deformation, whether elastic or permanent, may be of little importance as a design criterion.

2.3.2 Control of first surface yield or initiation of fracture

If it is desired to avoid permanent set in a plate panel, as a limit of serviceability, material yielding should not take place anywhere in the plate under working conditions. For a ship deck such as shown in Fig. 2.8 under wheel loading from a fork-lift truck, local yielding under repeated passages of the truck may result in undesirable excessive deformations in the form of dishing of the plates between the stiffeners, and should be avoided.

The criterion for the initiation of yielding in a ductile material is a subject yet unresolved for the general case. There are several different postulations put forward as to under what state of stress an element of a ductile material yields (Feodosyev 1968, Fluegge 1962). For steel sheets, the widely used criterion with a strong experimental support is that suggested by von Mises.

To use the yield criterion, first the biaxial state of stress in the element of plate under consideration is converted into an equivalent uniaxial stress σ_e, using the following relationship proposed by von Mises:

$$\sigma_e = (\sigma_x^2 + \sigma_y^2 - \sigma_x\sigma_y + 3\tau^2)^{\frac{1}{2}} \qquad (2.1)$$

The element of the plate is considered to yield when the magnitude of the equivalent stress thus obtained reaches the yield stress of the material in uniaxial tension, that is

$$\sigma_e = \sigma_y \qquad (2.2)$$

Maximum equivalent stress $\sigma_{e\,max}$ occurs at either the upper or the lower surface of the plate. It is not always immediately apparent at what location the maximum equivalent stress will occur, although it is usually either near the centre of the plate, or at the corner, or at the mid-point of the longer edges. The location may change as the load level increases. The results given in Chapter 4 are the maximum values, obtained by checking all over each surface for each load case.

For brittle materials, such as glass, fracture occurs at the location of the maximum tensile stress on the plate when this stress reaches the tensile strength of the material (Frownfelter, 1959). For the design of brittle sheets, a limit of unserviceability is the attainment of a maximum tensile stress σ_{max} equal to the statistically obtained fracture stress of the material divided by a factor of safety.

2.3.3 Allowance for limited spread of yield through the thickness

Ductile plates under transverse loading possess a great reserve of strength after first yield. Increase of loading after the onset of plasticity results in a spreading of yielding over the surface and its penetration through the thickness. Excessive local permanent deformation is limited if yielding does not spread through the whole thickness of the plate. Limited overall permanent set of a plate panel, such as permanent dishing of the bottom shell plate panels of a ship (Fig. 2.8) may be acceptable. In such a case greater economy is achieved through the selection of a thinner plate in which, under working conditions, local surface yielding is allowed to take place and penetrate through the thickness to a limited depth.

The penetration of plasticity through a given depth may not in itself be considered as a criterion without further qualification: rather the spread of the plastic zone over the plate, and more importantly the overall permanent deformation of the whole panel, should be the controlling design quantities. However, the aforementioned design quantities cannot be assessed without reverting to complicated elasto-plastic analyses. As a design procedure, therefore, the postulation of limited penetration of yielding through the thickness may be considered acceptable until a more exact criterion, amenable to design procedures, becomes available.

The depth of penetration may be approximately evaluated in terms of an elastic overstress of the maximum equivalent surface stress. It should be noted that the relationships between the depth of yield in a two-dimensionally stressed element of plate and the elastic surface over-stress are not simple. They depend on the magnitude and variation of individual stress components which are shown in Fig. 1.5. However, as a simplified approximate design guide assume that the two dimensional state of stress of an element of plate may be considered as equivalent to simple uniaxial bending with the maximum elastic stress given by $\sigma_{e\,max} = \sigma_y$ at the yielded surface as shown in Fig. 2.11(i), where α is the coefficient of elastic overstress and β is the coefficient expressing the depth of penetration. From the equilibrium of the resisting moments of the two distributions in Fig. 2.11 the following relationship may be deduced:

$$\alpha = 1 + 2\beta - 2\beta^2 \qquad (2.3)$$

From this relationship an over-stress of 10% corresponds to a depth of yield through the thickness equal to 5·3%, and a depth of yield of 25% from both surfaces corresponds to a surface over-stress of 37%. More exact relationships may be obtained if a uniform membrane stress is added to the distribution of stresses assumed in Fig. 2.11. Depending on the magnitude of the added membrane stress in relation to the maximum bending stress, similar relationships to (2.3) may be obtained, which result in different coefficients of over-stress and depth of penetration. The reader interested in the corresponding accurate but complex relationships in which the biaxial state of stress is allowed for is recommended to consult a recent investigation conducted by Burgoyne (1972).

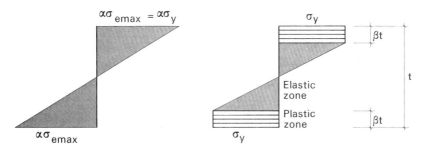

(i) Elastic distribution of stress with overstressing

(ii) Equivalent elasto-plastic distribution of stress with plasticity penetrating through depth βt

Fig. 2.11. Distribution of uniaxial bending stress through the thickness of an element of plate in an over-stressed linear elastic condition and its equivalent elasto-plastic distribution.

In the foregoing discussion it is assumed that the plate is predominantly acted upon by a transverse loading. When in-plane loading is also present, clearly the out-of-plane deformations become more important, and their influence with regard to the buckling strength of the plate should be critically examined.

With respect to control of yielding, the designer should bear in mind that, for panels of plating under frequently repeated loading, when fatigue becomes a design consideration, the stresses should be kept at low levels to comply with the fatigue life envisaged.

2.3.4 Control of mid-plane yielding of plate

In certain structural plates under transverse loading, considerations such as the possible action of a simultaneous in-plane loading, resistance to motion, maintenance, aesthetics and fatigue are not of design significance, as is the case for the bottom plates of the floating platform of Fig. 2.7, where excessive deformation of the skin plates may be allowed under working conditions. In this case, the mid-plane yielding of the plate may be defined as a limit of unserviceability.

The conditions resulting to the yielding of an element of plate through its thickness are discussed by several investigators (Illyushin 1956, Burgoyne 1972). These conditions are expressed in terms of the sectional actions, as listed in Fig. 1.4, which act on the element. Because the design procedure introduced in this volume is based on the maximum equivalent surface stress, not all the sectional actions necessary for the calculation of mid-plane yielding are given. It is therefore recommended that the approximate expression (2.3) with a value of yield penetration β equal to 0.5 be used; this corresponds to an over-stress of the elastic maximum equivalent stress of 50%.

At this stage, the attention of the designer is drawn to the fact that the elastic load–deflection and load–stress variations given herein are not strictly true for post yield loadings. However, recent research in progress (Lin *et al.*, 1972) suggest that as long as plastic zones do not penetrate fully through the thickness, that is to say the plastic zone being contained within a predominantly elastic region, the load–deflection and load–stress behaviour of the plate over the elastic zones remain close enough to the values given by the elastic theories. On this basis, the use of the elastic data given herein may be justified for design purposes in controlling the mid-plane yielding of the plate.

CHAPTER THREE
Design Parameters, Design Procedure, and Description of Typical Plate Problems Presented

3.1 Design parameters

At this stage, having gone through Chapters 1 and 2, the geometry of the plate (sides b and a, aspect ratio b/a) and its idealised boundary conditions may be assumed to have been determined. Before dealing with the next step in design, in the following, the non-dimensional parameters to be used in conjunction with the design curves are briefly described. The parameters given apply to both the rectangular and the circular plates.

(1) Transverse loading

For uniformly distributed loading the non-dimensional design parameter Q is expressed in terms of the transverse pressure q by the relationship

$$Q = a^4 q / t^4 E \tag{3.1}$$

For linearly varying distributions of pressure, such as container plates under hydrostatic pressure, the design parameter Q is given in terms of the maximum pressure q_{max}

$$Q = a^4 q_{max} / t^4 E \tag{3.2}$$

For concentrated loading p distributed uniformly over the patch area $u \times v$,

$$P = a^2 p / t^4 E \tag{3.3}$$

(?) Out-of-plane deflections

Transverse deflection w is given by

$$W = w/t \tag{3.4}$$

Numerical suffixes refer to the location on the plate.

(3) Maximum equivalent stress

Maximum equivalent stress anywhere in the plate is given by the non-dimensional parameter $\bar{\sigma}_{e\,max}$, which is related to the actual maximum equivalent stress $\sigma_{e\,max}$ by the expression

$$\bar{\sigma}_{e\,max} = (a/t)^2 \sigma_{e\,max}/E \tag{3.5}$$

At any value of loading Q or P, the equivalent stresses are checked all over the plate using the relationship (2.1). The maximum value thus obtained is given in the design curves.

(4) Maximum tensile stress

The non-dimensional maximum tensile stress $\bar{\sigma}_{max}$ is related to the actual maximum tensile stress σ_{max} by the following relationship:

$$\bar{\sigma}_{max} = (a/t)^2 \sigma_{max}/E \tag{3.6}$$

As in the preceding case, at each value of transverse loading, maximum tensile stresses are checked all over the plate using the relationship (3.7) and the maximum value thus obtained is recorded in the design curves given in Chapter 4.

$$\sigma_{max} = 0{\cdot}5(\sigma_x + \sigma_y) + [(\sigma_x - \sigma_y)^2/4 - \tau^2]^{\frac{1}{2}} \tag{3.7}$$

3.2 Design procedure

The next step in design usually poses itself in one of the following forms:

(1) For a given value of plate thickness t, it is required to check the deflection and stresses resulting under the action of transverse loading. Using the given t, calculate the non-dimensional transverse loading Q or P from the relationships (3.1)–(3.3).

Refer to the design curves, and for the value of Q or P calculated, read off the corresponding values of the non-dimensional deflection and stress coefficients (W, $\bar{\sigma}_{e\,max}$, $\bar{\sigma}_{max}$). Calculate the actual deflection and stresses as follows:

$$w = tW \text{ mm (in)} \tag{3.8}$$

$$\sigma_{e\,max} = E(t/a)^2 \bar{\sigma}_{e\,max} \text{ N/mm}^2 \text{ (psi)} \tag{3.9}$$

$$\sigma_{max} = E(t/a)^2 \bar{\sigma}_{max} \text{ N/mm}^2 \text{ (psi)} \tag{3.10}$$

(2) To maintain the out-of-plane deflections at a certain limit, and/or to keep the stresses below a permissible value, it is required to determine a suitable plate thickness for normal loading conditions q or p.

Assume a trial t. Knowing the values of transverse loading, a, E and the assumed value of t, calculate the non-dimensional value of the parameter controlling the design. For example, if maximum equivalent stress is to be kept at $\sigma_{e\,max}$, the parameter $\bar{\sigma}_{e\,max}$ is to be calculated. From the relating $\bar{\sigma}_{e\,max}/Q$ design curve in Chapter 4, read off the corresponding Q. Compare this Q with the value obtained from expressions (3.1)–(3.3), using the assumed t. For the correct value of t the two Q's should match. To be on the safe side, decide upon a value of t which results in the first Q being slightly greater than the second Q obtained from expressions (3.1)–(3.3). Now, with the Q thus obtained, refer to the W/Q design curves, from which the corresponding W is obtained, and hence

$$w = tW \text{ mm (in)}$$

If the actual deflection w, or the similarly obtained value of the actual maximum stress is not satisfactory, repeat the procedure with a new value of plate thickness t until acceptable limits for deflections and stresses are reached.

3.3 Poisson's ratio

The design curves are all for a Poisson's ratio of 0·3. For materials with values of Poisson's ratio other than 0·3, the corresponding design quantities are to be modified using the following relationships, in which a symbol marked with a star relates to the actual Poisson's ratio, v^*. For deflections

$$W^* = 1 \cdot 1(1 - v^{*2})W \qquad (3.11)$$

For bending stresses

$$\sigma_{bx}^* = 1 \cdot 1 \,|\, (1 - 0 \cdot 3v^*)\sigma_{bx} + (v^* - 0 \cdot 3)\sigma_{by}|$$
$$\sigma_{by}^* = 1 \cdot 1 \,|\, (1 - 0 \cdot 3v^*)\sigma_{by} + (v^* - 0 \cdot 3)\sigma_{bx}| \qquad (3.12)$$

For membrane stresses

$$\sigma_{mx}^* \simeq \sigma_{mx}$$
$$\sigma_{my}^* \simeq \sigma_{my} \qquad (3.13)$$

For the maximum equivalent and the maximum tensile stresses given, it is not possible to evaluate their corresponding accurate values for Poisson's ratio other than 0·3. However, an upper limit to the maximum deviation from the given maximum equivalent and maximum tensile stresses may be obtained by making the following assumptions:

(1) The contribution of membrane stresses is zero.
(2) For values of Poisson's ratio less than 0·3, the stresses are taken as uniaxial, and for Poisson's ratio greater than 0·3 the state of stress is biaxial with stresses equal in both directions.

The above assumptions are not of practical significance, but are required for the evaluation of the maximum possible values of $\sigma_{e\,max}$ and σ_{max} for Poisson's ratios different from 0·3. Thus

for $v^* < 0 \cdot 3$

$$\sigma_{e\,max}^* = 1 \cdot 1(1 - 0 \cdot 3v^*)\sigma_{e\,max}$$
$$\sigma_{max}^* = 1 \cdot 1(1 - 0 \cdot 3v^*)\sigma_{max} \qquad (3.14)$$

Design Parameters, Design Procedure; Typical Plate Problems 41

for $v^* > 0.3$

$$\sigma^*_{e\,\max} = 0.77(1 + v^*)\sigma_{e\,\max}$$
$$\sigma^*_{\max} = 0.77(1 + v^*)\sigma_{\max}$$
(3.15)

The expressions (3.14), (3.15) may be used as a guide in design. The actual values of stresses for a Poisson's ratio other than 0·3 are in fact closer to the values given in the curves than the adjusted values obtained from the preceding examples.

3.4 List of plate problems treated

The plate problems for which design data are provided are grouped into five categories and are listed in Figs. 3.1–3.5. Each figure, referring to one of the categories, shows the related boundary conditions and loading as well as the code number of the design curves in Chapter 4 for each of the cases treated.

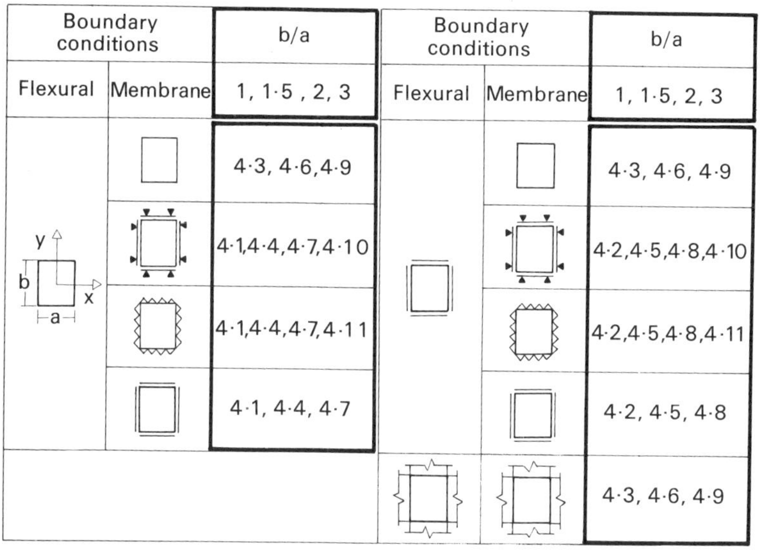

Boundary conditions		b/a	Boundary conditions		b/a
Flexural	Membrane	1, 1·5, 2, 3	Flexural	Membrane	1, 1·5, 2, 3
□		4·3, 4·6, 4·9	□		4·3, 4·6, 4·9
	▢	4·1, 4·4, 4·7, 4·10		▢	4·2, 4·5, 4·8, 4·10
	▨	4·1, 4·4, 4·7, 4·11		▨	4·2, 4·5, 4·8, 4·11
	□	4·1, 4·4, 4·7		□	4·2, 4·5, 4·8
					4·3, 4·6, 4·9

Fig. 3.1. Rectangular plates with symmetrical boundary conditions under uniform transverse pressure. Code number of design curves for different boundary conditions and aspect ratios.

Boundary conditions		b/a	α×β	Boundary conditions		b/a	α×β
Flexural	Membrane		0.1x0.1 0.2x0.2 0.3x0.3 0.2x0.3 0.2x0.4	Flexural	Membrane		0.1x0.1 0.2x0.2 0.3x0.3 0.2x0.3 0.2x0.4
		1	4.25			1	4.33
		1.5	4.26				
		2	4.27			2	4.34
		3	4.28				
		1	4.29, 4.37			1	4.35
		1.5	4.30, 4.37				
		2	4.31, 4.38			2	4.36
		3	4.32, 4.38				

Fig. 3.2. Rectangular plates under central concentrated loading. Code number of design curves for different boundary conditions, aspect ratios and patch dimensions (dimensions of loaded area).

Boundary conditions		b/a			Boundary conditions		b/a		
Flexural	Membrane	1	1.5	2	Flexural	Membrane	1	1.5	2
		4.39					4.44		
							4.45		
		4.40							
		4.41					4.46		
		4.42, 4.43					4.47		
		4.41							
		4.42, 4.43							

Fig. 3.3. Rectangular plates with unsymmetrical boundary conditions under uniform distributed loading. Code number of design curves for different boundary conditions and aspect ratios.

Design Parameters, Design Procedure; Typical Plate Problems 43

Boundary conditions		b/a				Boundary conditions		b/a			
Flexural	Membrane	0·67	1	1·5	2	Flexural	Membrane	0·67	1	1·5	2
(y,b,a,x diagram)	□		4·48				□		4·52		
	▶□◀		4·49				▶□◀		4·53		
[ ]	□		4·50			[ ]	□		4·54		
	▶□◀		4·51				▶□◀		4·55		

Fig. 3.4. Rectangular plates under a linearly varying distribution of transverse pressure (hydrostatic pressure). Code number of design curves for different boundary conditions and aspect ratios.

Boundary conditions		u/a	Boundary conditions		u/a
Flexural	Membrane	0·05, 0·10 0.15, 1	Flexural	Membrane	0·05, 0·10 0·15. 1
(circle with a)	○	4·58	(circle with hatched center)	○	4·60
	○ (arrows)	4·59		○ (arrows)	4·61

Fig. 3.5. Circular plates under central concentrated loadings. Code number of design curves for different boundary conditions and patch dimensions.

CHAPTER FOUR
Design Curves and Data

4.1 Symmetrical plates under uniform transverse loading

4.1.1 Design curves

As discussed in Chapter 2, the design of a thin plate under transverse loading may be reduced to the control of out-of-plane deflections w, maximum equivalent surface stress $\sigma_{e\,max}$, and the maximum surface tensile stress σ_{max}.

For increasing values of transverse pressure Q, values of W, $\bar{\sigma}_{e\,max}$ and $\bar{\sigma}_{max}$ are given in Figs. 4.1–4.9 for the boundary conditions and aspect ratios listed in Fig. 3.1. Where the plates are restrained against in-plane movement, the membrane forces developed at the edges are given for increasing values of Q (Figs. 4.10, 4.11) so as to enable the forces required for the design of boundary welds or rivets to be evaluated. Numerical examples for the use of the design curves are given in Chapter 5.

For simply supported plates with no rotational or membrane restraint, i.e. truly free conditions, it is recommended that the design curves given at the top of Figs. 4.1, 4.4, 4.7 are used. At the other extreme, for plates with maximum restraint at the edges, the bottom curves of Figs. 4.2, 4.5, 4.8 are to be used. Note that in the latter case full fixity against tangential movement at the boundaries is not satisfied. However, as it is shown in section 4.1.2, the specified boundary conditions may be assumed to represent full fixity of the supports at the edges for design purposes.

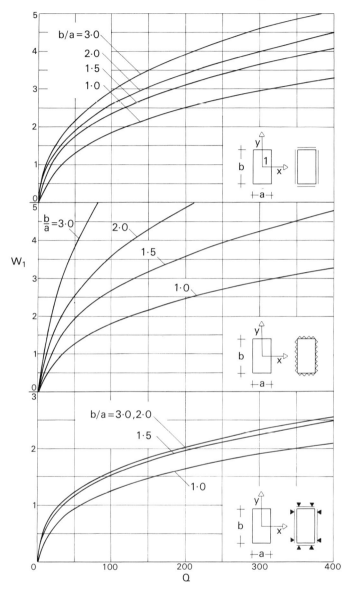

Fig. 4.1. Symmetrical plates under uniform transverse loading. Variations of central deflection W with intensity of transverse pressure Q.

46 Thin Plate Design For Transverse Loading

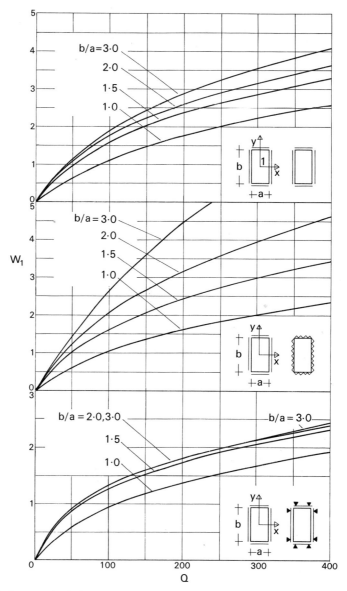

Fig. 4.2. Symmetrical plates under uniform transverse loading. Variations of central deflection W with intensity of transverse pressure Q.

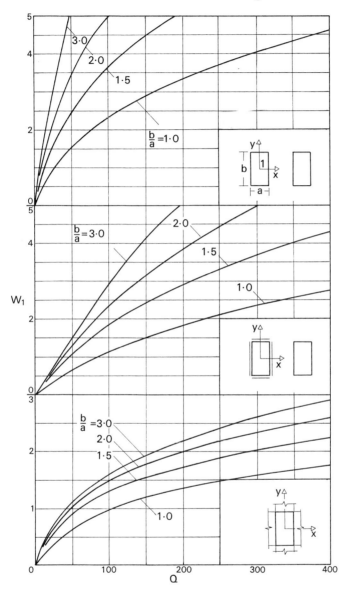

Fig. 4.3. Symmetrical plates under uniform transverse loading. Variations of central deflection W with intensity of transverse pressure Q.

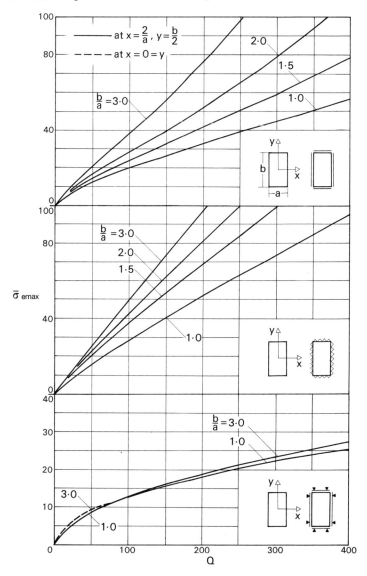

Fig. 4.4. Symmetrical plates under uniform transverse loading. Variations of maximum equivalent surface stress $\bar{\sigma}_{e\,max}$ with transverse loading Q.

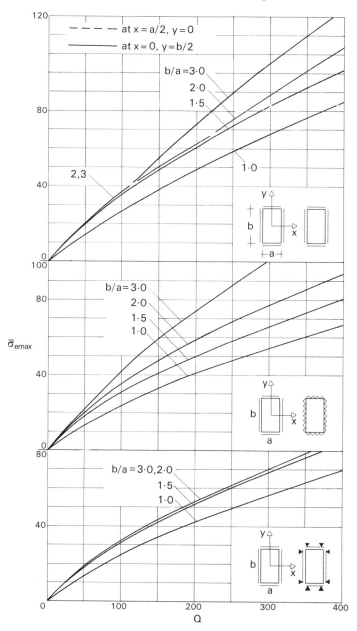

Fig. 4.5. Symmetrical plates under uniform transverse pressure. Variations of maximum equivalent surface stress $\bar{\sigma}_{e\,\text{max}}$ with transverse loading Q.

50 Thin Plate Design For Transverse Loading

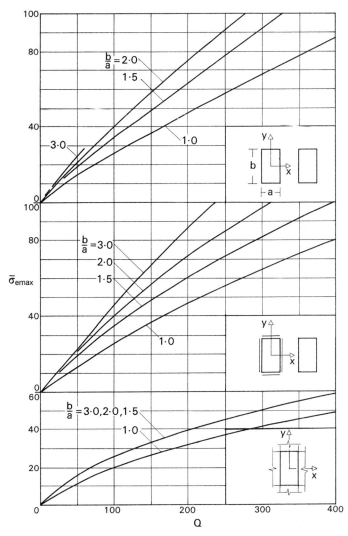

Fig. 4.6. Symmetrical plates under uniform transverse pressure. Variations of maximum equivalent surface stress $\bar{\sigma}_{e\,\max}$ with transverse loading Q.

Design Curves and Data 51

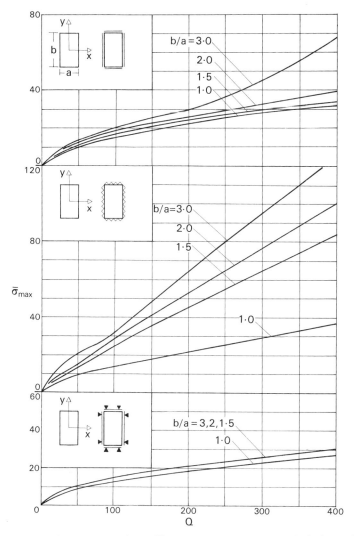

Fig. 4.7. Symmetrical plates under uniform transverse pressure. Variations of maximum tensile surface stress $\bar{\sigma}_{max}$ with transverse loading Q.

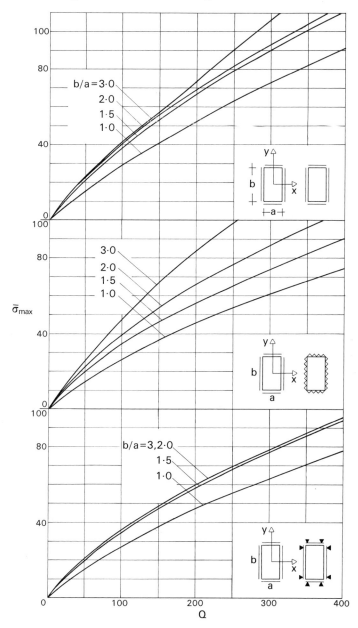

Fig. 4.8. Symmetrical plates under uniform transverse pressure. Variations of maximum tensile surface stress $\bar{\sigma}_{max}$ with transverse loading Q.

Design Curves and Data 53

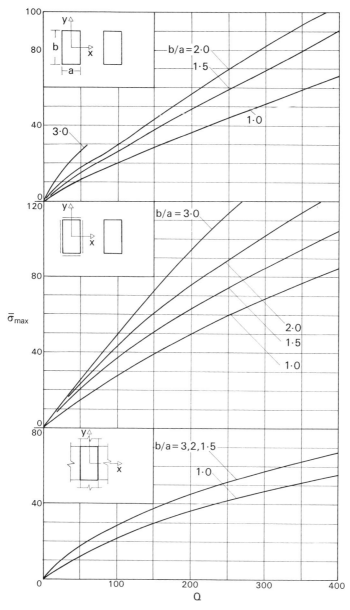

Fig. 4.9. Symmetrical plates under uniform transverse pressure. Variations of maximum tensile surface stress $\bar{\sigma}_{max}$ with transverse loading Q.

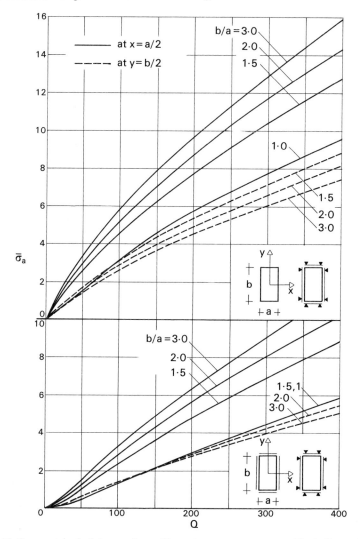

Fig. 4.10. Symmetrical plates under uniform transverse pressure. Variations of the average tensile membrane stresses $\bar{\sigma}_a$ developed at the boundaries with intensity of transverse loading Q.

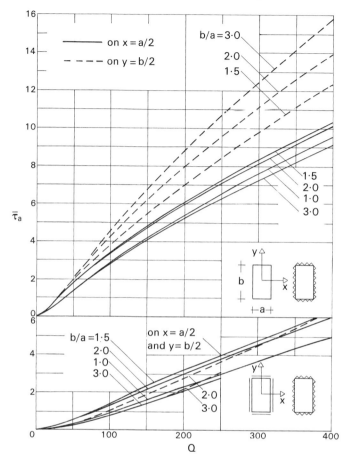

Fig. 4.11. Symmetrical plates under uniform transverse pressure. Variations of average shearing membrane stresses $\bar{\tau}_a$ developed at the boundaries with intensity of transverse loading Q.

4.1.2 Detailed description of large-deflection behaviour

To provide the reader with a deeper insight into the large-deflection behaviour of plates, in the following section the influences of the membrane boundary conditions and some of the more significant stress distributions are briefly discussed.

(1) Influences of membrane boundary conditions

The distributions of bending and membrane stresses across the centre lines and the boundaries of a rotationally free rectangular plate with aspect ratio 2 are shown in Fig. 4.12 for the four different limiting membrane boundary conditions considered. The figure illustrates the widely different membrane stress distributions resulting from differences assumed in the membrane boundary conditions. Similar distributions for the rotationally fixed plates for different membrane boundary conditions are shown in Fig. 4.13. Note that in the latter case the intensity of transverse pressure Q is the same for the four cases shown. The differences between the magnitudes of central deflections obtained (about 150%) is due to the differences in the membrane boundary conditions.

In both figures, on account of equilibrium considerations, stresses $\bar{\sigma}_{my}$ across $y = 0$ and $\bar{\sigma}_{mx}$ across $x = 0$ are each in self equilibrium for the cases (i). For the cases (ii), $\bar{\sigma}_{mx}$ at $x = 0$ is in equilibrium with $\bar{\sigma}_{mx}$ at $x = a/2$. Similarly, $\bar{\sigma}_{my}$ at $y = 0$ and $\bar{\sigma}_{my}$ at $y = b/2$ are in static equilibrium with one another. For the plates restrained against tangential movement at the boundaries, i.e. cases (iii), the boundary membrane shear stresses developed (τ_m) at $x = a/2$ are in equilibrium with $\bar{\sigma}_{my}$ at $y = 0$. Likewise, τ_m at $y = b/2$ is in equilibrium with $\bar{\sigma}_{mx}$ at $x = 0$. The membrane stresses of cases (iv) across the boundaries and the centre lines are each in self equilibrium.

In sections (v) of both figures the corresponding distributions of small-deflection bending stresses are shown.

(2) Large-deflection distributions of bending stresses

An increase in the out-of-plane deflections is accompanied by a change in the distributions of bending moments throughout the plate. The change, in general, is a reduction of relative curvatures (hence bending moments) at the centre of the plate, coupled with an increase in the relative curvatures near the boundaries. For the rotationally free rectangular plate of cases (i) and (ii) of Fig. 4.12, the corresponding bending stress profiles (proportional to bending moments) are shown in Fig. 4.14 to illustrate this point. The profiles shown are normalised with respect to their values at plate centre to afford easy comparison.

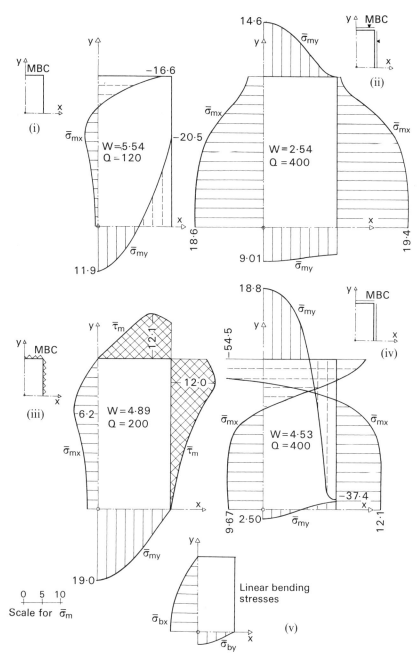

Fig. 4.12. Rotationally free rectangular plates ($b/a = 2$) under uniform transverse pressure. Distributions of membrane stress along the centre lines and the boundaries for different membrane boundary conditions.

58 Thin Plate Design For Transverse Loading

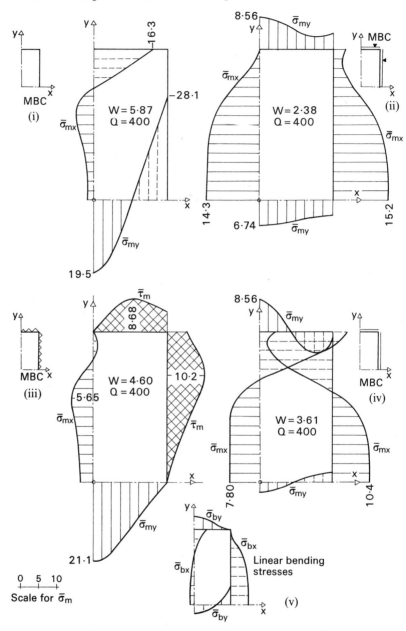

Fig. 4.13. Rotationally fixed rectangular plates ($b/a = 2$) under uniform transverse pressure. Distributions of membrane stress along the centre lines and the boundaries for different membrane boundary conditions and $Q = 400$.

The corresponding profiles from the small-deflection theory are also included to illustrate the differences.

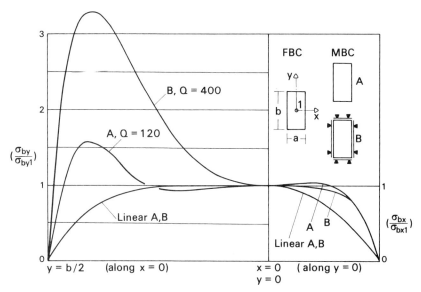

Fig. 4.14. Rotationally free rectangular plates ($b/a = 2$) under uniform transverse pressure. Bending-stress profiles along the centre lines for different membrane boundary conditions.

(3) Maximum equivalent stresses

At each point on the plate the maximum equivalent stress occurs at the plate surface, either at the loaded or the unloaded side. Typical distributions of equivalent stresses at both surfaces are shown in Fig. 4.15 for two rotationally free plates and in Fig. 4.16 for two rotationally fixed plates. The distributions show lines of equal equivalent stresses over both the loaded and unloaded surfaces. The figures are representative of the majority of rotationally free and fixed plates, where in the former group the maximum equivalent stress throughout the plate occurs near the corners and in the latter group at the midpoint of the longer edges.

(4) Maximum tensile stress

For simply supported plates the maximum tensile stress occurs at the unloaded surface, either near the plate centre or the plate corner,

60 Thin Plate Design For Transverse Loading

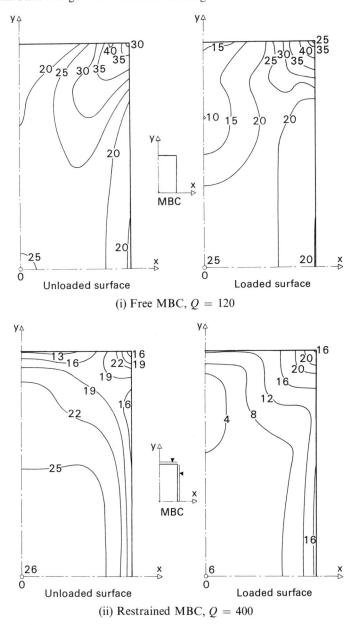

Fig. 4.15. Rotationally free rectangular plates ($b/a = 2$) under uniform transverse pressure. Contours of equivalent surface stresses $\bar{\sigma}_e$ at the loaded and unloaded surfaces for different membrane boundary conditions.

Design Curves and Data 61

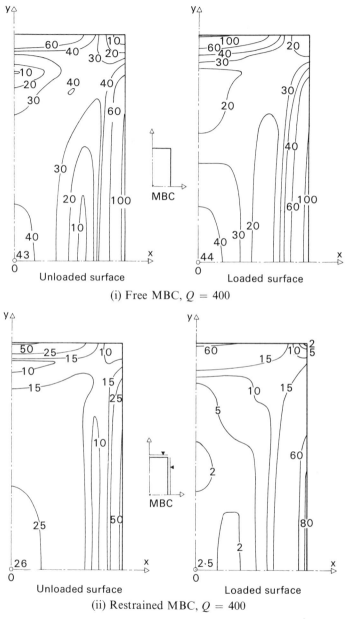

Fig. 4.16. Rotationally fixed rectangular plates ($b/a = 2$) under uniform transverse pressure. Contours of the equivalent surface stresses $\bar{\sigma}_e$ at the loaded and unloaded surfaces for different membrane boundary conditions.

62 Thin Plate Design For Transverse Loading

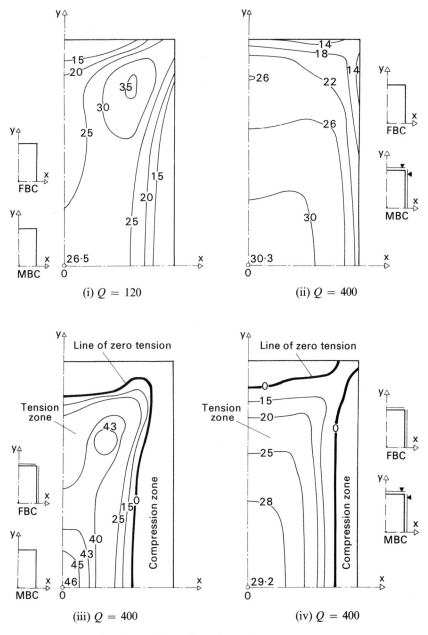

Fig. 4.17. Rectangular plates ($b/a = 2$) under uniform transverse pressure. Contours of the maximum tensile surface stresses $\bar{\sigma}_{max}$ at the unloaded surface.

depending on the membrane boundary conditions as shown in Fig. 4.17 (i) and (ii). For the rotationally fixed plates, at low values of transverse pressure, maximum tensile stress occurs at the mid-point of the longer sides on the loaded surface, but with increase in the transverse deflections and the development of membrane stresses, the position of the maximum tensile stress may shift to the plate centre depending on the membrane boundary conditions.

(5) Conditions of full fixity

Throughout this book, the boundary conditions of rotational fixity plus zero extensional displacement of the edges coupled with zero membrane shear stress as shown symbolically in Fig. 4.18(i) is assumed to represent the condition of full fixity of the edges. However, strictly speaking, for full fixity the tangential displacement at the boundaries should be equated to zero, as shown symbolically in Fig. 4.18(ii) instead of the membrane shear stresses. The former idealisation is adopted here, because the numerical values of the design quantities for the two cases do not differ significantly, with the difference that the former idealisation is simpler to apply.

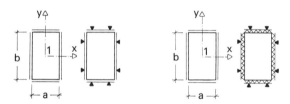

(i) Assumed idealisation (ii) Strictly true idealisation

Fig. 4.18. Fully fixed boundary-condition idealisations.

For example in the case of a square plate acted upon by a transverse pressure $Q = 400$, the results of the analyses show that the simpler idealisation adopted (Fig. 4.18(i)) over-estimates the design quantities on the safe side by the following percentages.

Central deflection W_1 0.79%
Maximum equivalent stress $\bar{\sigma}_{e\,max}$ 2.19%
Maximum tensile stress $\bar{\sigma}_{max}$ 1.30%
Average reactive membrane direct stress at the boundaries $\bar{\sigma}_a$ 14.6%

The average reactive membrane shear stresses ($\bar{\tau}_a$) developed at the boundaries of the idealisation of Fig. 4.18(ii) are about 20·1% of the corresponding average direct stresses quoted, that is

$$\bar{\tau}_a \simeq 0{\cdot}2\bar{\sigma}_a \tag{4.1}$$

4.1.3 Approximate solutions and plates with infinite aspect ratios

(1) Plates with finite aspect ratios

For rectangular plates with finite aspect ratios under uniform transverse pressure there are several approximate solutions proposed, some of which are reported in the references (Eggwertz and Norr 1953, Bares 1971, Liptak et al. 1973).

(a) *Simply supported plates with free membrane boundary conditions* (Fig. 4.19)

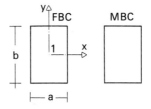

Fig. 4.19. Rectangular plates under uniform transverse pressure.

The approximate expressions derived by Bares (1971) for deflections and stresses of a rectangular plate with aspect ratio $\rho = b/a$ and Poisson's ratio of 0·3 can be expressed in terms of the present notation in the following form.

$$\frac{19{\cdot}43}{(1 + 0{\cdot}6\rho^2 + \rho^4)} W_1^3 + 5{\cdot}5(1 + 1/\rho^2)W_1 = Q \tag{4.2}$$

$$\bar{\sigma}_{bx1} = 5{\cdot}423(1 + 0{\cdot}3/\rho^2)W_1 \tag{4.3}$$

$$\bar{\sigma}_{by1} = 5{\cdot}423(1/\rho^2 + 0{\cdot}3)W_1 \tag{4.4}$$

$$\bar{\sigma}_{mx1} = \frac{2{\cdot}76}{(1/\rho^2 + 0{\cdot}6 + \rho^2)} W_1^2 \tag{4.5}$$

$$\bar{\sigma}_{my1} = \frac{2{\cdot}76}{(1 + 0{\cdot}6\rho^2 + \rho^4)} W_1^2 \tag{4.6}$$

For a plate with aspect ratio $\rho = 2$ and $Q = 400$, the values obtained from the approximate relationships (4.2)–(4.6) and the corresponding values from the present work are given for comparison in Table 4.1. The results of this approximate analysis do not appear to be accurate enough for design purposes.

Table 4.1

	Approximate analysis	Present analysis
W_1	6.98	9·79
$\bar{\sigma}_{bx1}$	40·7	35·3
$\bar{\sigma}_{by1}$	20·8	17·1
$\bar{\sigma}_{mx1}$	27·7	2·37
$\bar{\sigma}_{my1}$	6·92	31·3

(b) *Plates with restrained membrane boundary conditions* (Fig. 4.20)

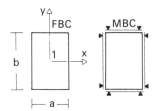

Fig. 4.20. Rectangular plates with restrained membrane boundary conditions under uniform transverse pressure.

Similarly for rotationally free plates with restrained membrane boundary conditions, the corresponding relationships are:

$$(12 + 12/\rho^4 + 4·95/\rho^2)W_1^3 + 5·5(1 + 1/\rho^2)^2 W_1 = Q \quad (4.7)$$

$$\bar{\sigma}_{bx1} = 5·42(1 + 0·3/\rho^2)W_1 \quad (4.8)$$

$$\bar{\sigma}_{by1} = 5·42(1/\rho^2 + 0·3)W_1 \quad (4.9)$$

$$\bar{\sigma}_{mx1} = 1·35(1·91 + 0·3/\rho^2)W_1^2 \quad (4.10)$$

$$\bar{\sigma}_{my1} = 1·35(1·91/\rho^2 + 0·3)W_1^2 \quad (4.11)$$

66 Thin Plate Design For Transverse Loading

It should be noted that in both the preceding cases, the maximum equivalent stress and the maximum tensile stress which are of design interest do not necessarily occur at plate centre, for which the approximate solutions apply.

(2) Infinitely long rectangular plates under uniform transverse pressure

(a) Rotationally free plates with the edges free to move in (Fig. 4.21)

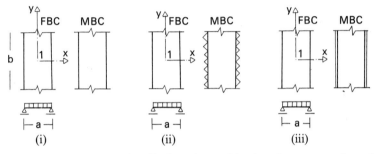

Fig. 4.21. Infinitely long rotationally free plates with edges free to move in under uniform transverse pressure.

Regions of the plate away from the ends undergo cylindrical bending in all three cases shown in Fig. 4.21. For this region, no membrane forces develop in the x-direction. The behaviour of the plate away from the ends is therefore similar to a simply supported beam with the values of deflections and stresses given by

$$W_1 = 0.142Q \tag{4.12}$$

$$\bar{\sigma}_{bx1} = 0.682Q \tag{4.13}$$

$$\bar{\sigma}_{by1} = -0.205Q \tag{4.14}$$

$$\bar{\sigma}_{mx1} = 0 \tag{4.15}$$

$$\bar{\sigma}_{my1} = 0 \quad \text{for cases (i) and (iii)} \tag{4.16}$$

$$\bar{\sigma}_{e\,max} = 0.804Q \tag{4.17}$$

$$\bar{\sigma}_{max} = 0.682Q \tag{4.18}$$

(b) *Rotationally fixed plates with the edges free to move in* (Fig. 4.22)

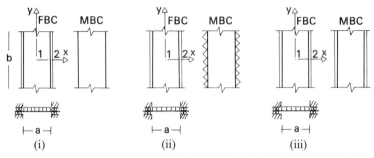

Fig. 4.22. Infinitely long rotationally fixed plates with edges free to move in under uniform transverse pressure.

In a similar way to the preceding cases, there is no large deflection behaviour for the plates shown in Fig. 4.22. The design quantities, derived from simple beam theory, are as follows:

$$W_1 = 0.0284Q \tag{4.19}$$

$$\bar{\sigma}_{bx2} = 0.455Q \tag{4.20}$$

$$\bar{\sigma}_{by2} = -0.1365Q \tag{4.21}$$

$$\bar{\sigma}_{mx2} = 0 \tag{4.22}$$

$$\bar{\sigma}_{my2} = 0 \tag{4.23}$$

$$\bar{\sigma}_{e\,max} = 0.536Q \tag{4.24}$$

$$\bar{\sigma}_{max} = 0.455Q \tag{4.25}$$

(c) *Rotationally free plates with the edges restrained against in-plane movement* (Fig. 4.23)

Using the relationships derived by Bares (1971), the design quantities may be expressed in terms of the present notation and a parameter k in the following forms:

$$Q = 2.58k(1 + 0.405k^2) \tag{4.26}$$

$$W_1 = 0.3412 \left[\frac{2(\operatorname{sech} k - 1) - k^2}{k^4} \right] Q \tag{4.27}$$

$$\bar{\sigma}_{bx1} = 1\cdot 5Q(1 - \text{sech } k)/k^2 \qquad (4.28)$$

$$\bar{\sigma}_{by1} = -0\cdot 3\bar{\sigma}_{bx1} \qquad (4.29)$$

$$\bar{\sigma}_{mx1} = \bar{\sigma}_{mx2} = 0\cdot 366k^2 \qquad (4.30)$$

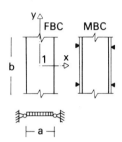

Fig. 4.23. Infinitely long rotationally free rectangular plates with edges restrained against in-plane movement under uniform transverse pressure.

(d) *Rotationally fixed plates with the edges restrained against in-plane movement* (Fig. 4.24)

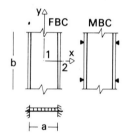

Fig. 4.24. Infinitely long rotationally fixed rectangular plates with edges restrained against in-plane movement under uniform transverse pressure.

The corresponding approximate design quantities are (Bares 1971)

$$Q = 12\cdot 92k(1 + 0\cdot 101k^2) \qquad (4.31)$$

$$W_1 = 0\cdot 005\,69(0\cdot 5k^2 + k/\sinh k - k/\tanh k)Q/k^2 \qquad (4.32)$$

$$\bar{\sigma}_{bx2} = -1\cdot 5Q(k - \tanh k)/k^2 \tanh k \qquad (4.33)$$

$$\bar{\sigma}_{by2} = -0\cdot 3\bar{\sigma}_{bx2} \qquad (4.34)$$

$$\bar{\sigma}_{mx2} = 0\cdot 366k^2 \qquad (4.35)$$

$$\bar{\sigma}_{my2} = 0\cdot 3\bar{\sigma}_{mx2} \qquad (4.36)$$

4.1.4 Related work by other investigators

Steinhardt and Abdel-Sayed (1963) have presented approximate solutions for continuous plates. Clarkson (1963) has carried out some experimental work on plates under uniform transverse pressure covering the range of small deflections to large-deflection elastoplastic region.

For the ultimate strength of plates Jones and Walters (1971) have used a rigid-plastic analysis to obtain large-deflection ultimate strength solutions of transversely loaded plates. Wah (1960) has also presented a treatment for the rigid-plastic ultimate strength of plates.

More work on the large deflection of plates under uniform transverse loading is reported by Williams (1955), Bauer *et al.* (1964), Scholes and Bernstein (1969).

4.2 Rectangular plates under central concentrated loadings

4.2.1 Design curves

Design curves are given for square and rectangular plates with aspect ratios 1–3 and boundary conditions covering the range of fully free to fully fixed conditions. The loading P is assumed to be acting at the centre of the plate and to be distributed uniformly (intensity p) over a rectangular patch area with dimensions $u \times v$. For each case, central patch loadings with five different dimensions are considered ($\alpha \times \beta$ being 0.1×0.1, 0.2×0.2, 0.3×0.3, 0.2×0.3, 0.2×0.4). A sixth condition of uniformly distributed loading over the whole area UDL is included in each case for comparison.

The design curves are presented such that for assumed values of concentrated loading P, and plate aspect ratio and boundary conditions, the corresponding central deflections W, maximum equivalent stress $\sigma_{e\,max}$, and the maximum tensile stress σ_{max} may be readily read off from the curves for a range of patch dimensions.

As it may be inferred from the design curves, for deflections it is unnecessary to have an accurate knowledge of the wheel contact area. It should also be noted that for higher values of loading and small patch sizes (e.g. $\alpha \times \beta = 0.1 \times 0.1$), the magnitudes of bending and membrane stresses for the two extreme cases of simply supported plates (Figs. 4.25–4.28) and fully fixed conditions (Figs. 4.29–4.32)

70 Thin Plate Design For Transverse Loading

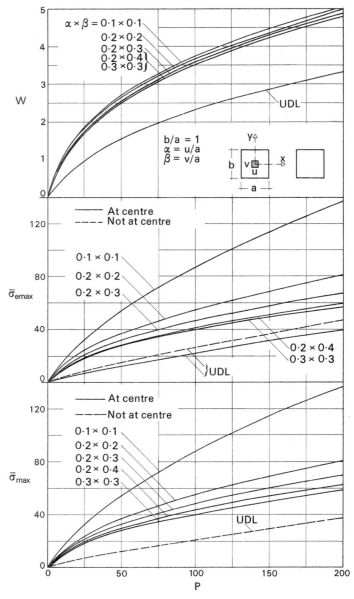

Fig. 4.25. Rectangular plates under a central concentrated loading P. Variations with P of central deflection W, maximum equivalent stress $\bar{\sigma}_{e\,\text{max}}$, and maximum tensile stress $\bar{\sigma}_{\text{max}}$.

Design Curves and Data 71

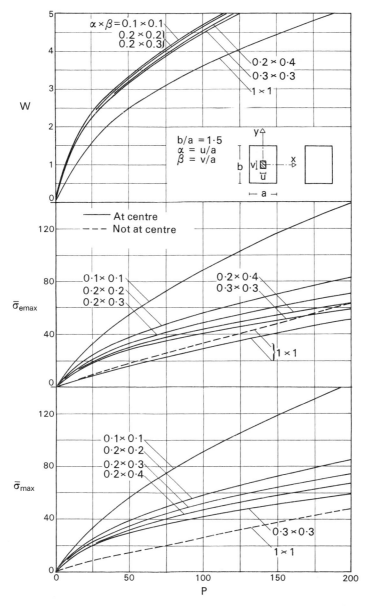

Fig. 4.26. Rectangular plates under a central concentrated loading P. Variations with P of central deflection W, maximum equivalent stress $\bar{\sigma}_{e\,\max}$, and maximum tensile stress $\bar{\sigma}_{\max}$.

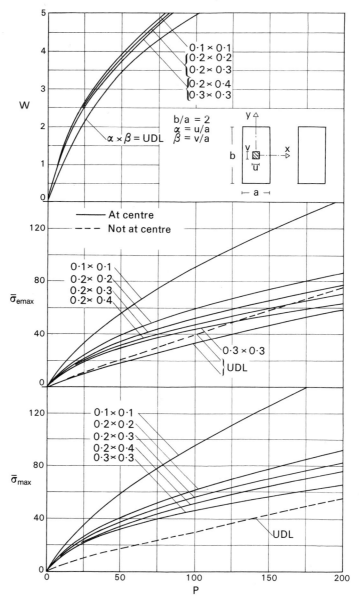

Fig. 4.27. Rectangular plates under a central concentrated loading P. Variations with P of central deflection W, maximum equivalent stress $\bar{\sigma}_{e\,max}$, and maximum tensile stress $\bar{\sigma}_{max}$.

Design Curves and Data 73

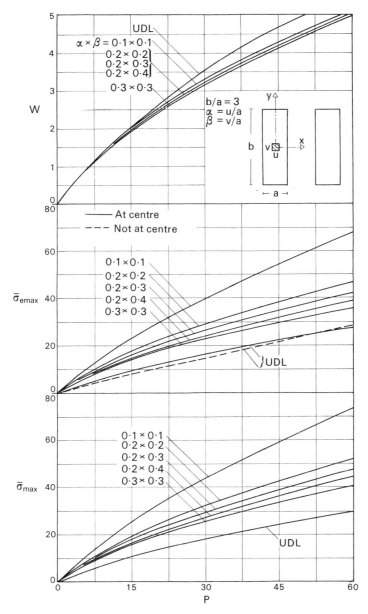

Fig. 4.28. Rectangular plates under a central concentrated loading P. Variations with P of central deflection W, maximum equivalent stress $\bar{\sigma}_{e\,\mathrm{max}}$, and maximum tensile stress $\bar{\sigma}_{\mathrm{max}}$.

74 Thin Plate Design For Transverse Loading

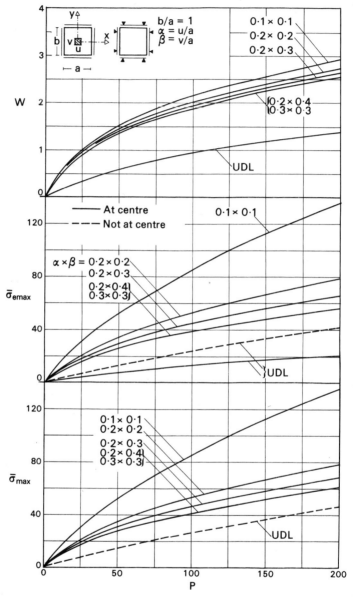

Fig. 4.29. Rectangular plates under a central concentrated loading P. Variations with P of central deflection W, maximum equivalent stress $\bar{\sigma}_{e\,max}$, and maximum tensile stress $\bar{\sigma}_{max}$.

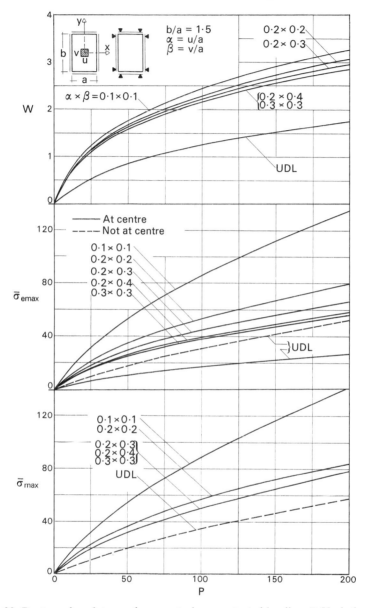

Fig. 4.30. Rectangular plates under a central concentrated loading P. Variations with P of central deflection W, maximum equivalent stress $\bar{\sigma}_{e\,max}$, and maximum tensile stress $\bar{\sigma}_{max}$.

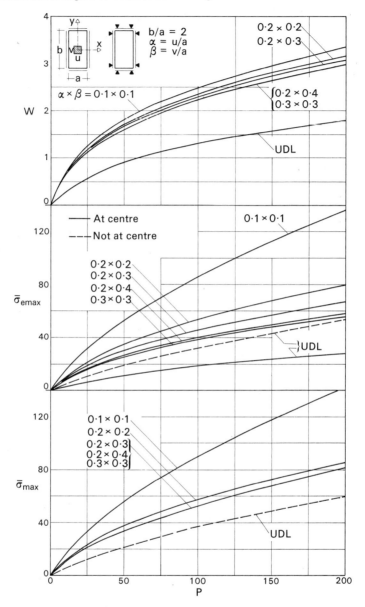

Fig. 4.31. Rectangular plates under a central concentrated loading P. Variations with P of central deflection W, maximum equivalent stress $\bar{\sigma}_{e\,max}$, and maximum tensile stress $\bar{\sigma}_{max}$.

Design Curves and Data 77

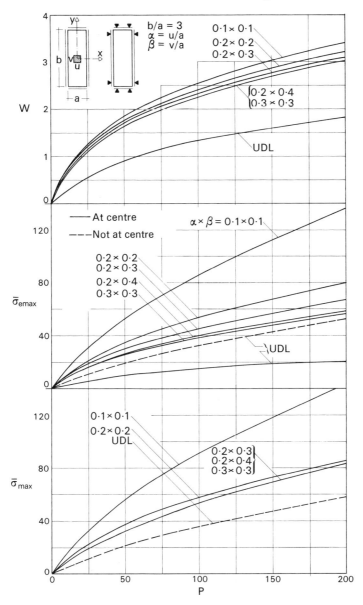

Fig. 4.32. Rectangular plates under a central concentrated loading P. Variations with P of central deflection W, maximum equivalent stress $\bar{\sigma}_{e\,\max}$, and maximum tensile stress $\bar{\sigma}_{\max}$.

78 Thin Plate Design For Transverse Loading

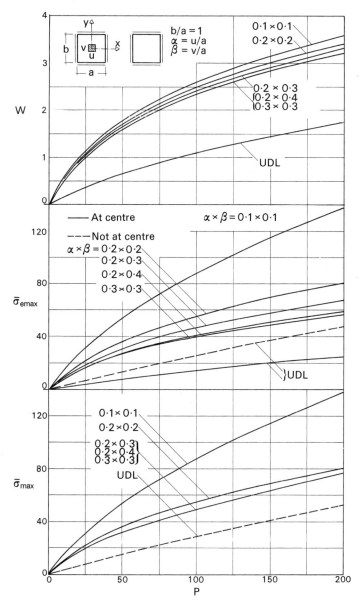

Fig. 4.33. Rectangular plates under a central concentrated loading P. Variations with P of central deflection W, maximum equivalent stress $\bar{\sigma}_{e\,max}$, and maximum tensile stress $\bar{\sigma}_{max}$.

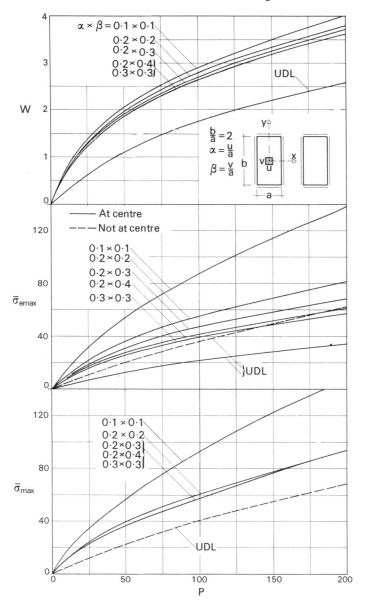

Fig. 4.34. Rectangular plates under a central concentrated loading P. Variations with P of central deflection W, maximum equivalent stress $\bar{\sigma}_{e\,\text{max}}$, and maximum tensile stress $\bar{\sigma}_{\text{max}}$.

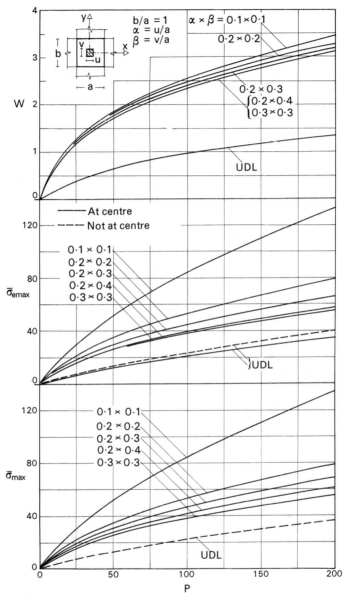

Fig. 4.35. Rectangular plates under a central concentrated loading P. Variations with P of central deflection W, maximum equivalent stress $\bar{\sigma}_{e\,max}$, and maximum tensile stress $\bar{\sigma}_{max}$.

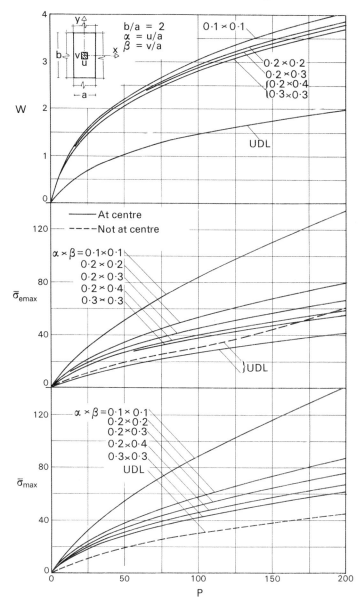

Fig. 4.36. Rectangular plates under a central concentrated loading P. Variations with P of central deflection W, maximum equivalent stress $\bar{\sigma}_{e\max}$, and maximum tensile stress $\bar{\sigma}_{\max}$.

82 Thin Plate Design For Transverse Loading

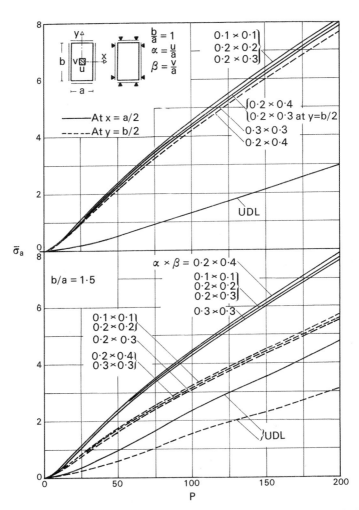

Fig. 4.37. Rectangular plates under a central concentrated loading P. Variations with P of the average reactive membrane direct stress $\bar{\sigma}_a$ developed at the boundaries.

are each within 3% of one another at centre. It may be concluded that at higher values of loading, the stresses under a patch loading are not influenced significantly by the plate's outer boundary conditions and a rigorous analysis for the evaluation of boundary restraints under

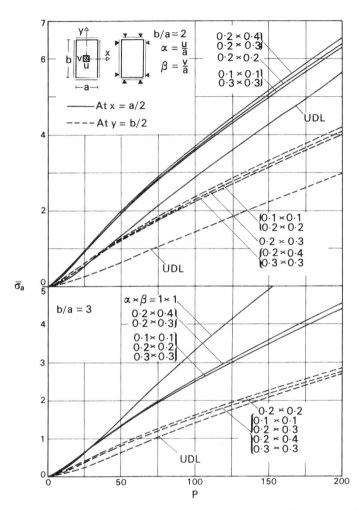

Fig. 4.38. Rectangular plates under a central concentrated loading P. Variations with P of the average reactive membrane direct stress $\bar{\sigma}_a$ developed at the boundaries.

practical conditions may not be justified. The two extreme solutions given predict with reasonable accuracy the final stresses in a practical plate.

4.2.2 Related work

Little work is done on the large-deflection elastic behaviour of plates under concentrated loading. Weiss (1969), using finite differences, has obtained solutions for continuous rectangular plates under a central patch loading. A detailed discussion on the large-deflection behaviour of plates under concentrated loading is given in Aalami (1973). Experimental results on steel plates acted upon by wheel loadings are reported by Smith (1963).

4.3 Rectangular plates with unsymmetrical boundary conditions under uniform transverse pressure

4.3.1 Design curves

Figures 4.39–4.47 give design curves for plates with unsymmetrical boundary conditions under uniform transverse pressure. For each specific set of boundary conditions the aspect ratios treated are $b/a = 0.5, 1, 1.5, 2$. For convenience, through appropriate change of the coordinate axes, these are arranged in groups with aspect ratios greater than 1, as may be observed from the design curves.

Design Curves and Data 85

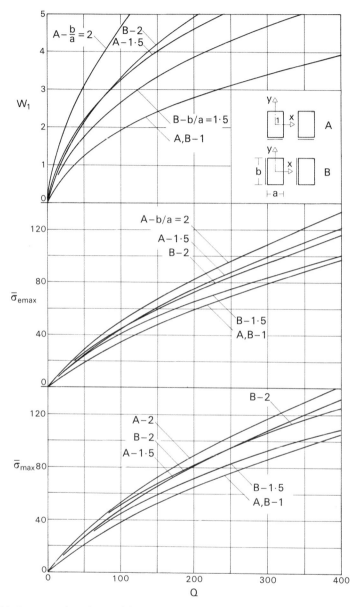

Fig. 4.39. Rectangular plates with unsymmetrical boundary conditions under uniform transverse pressure Q. Variations, with Q, of the transverse deflection W, maximum equivalent stress $\bar{\sigma}_{e\max}$, and the maximum stress $\bar{\sigma}_{\max}$.

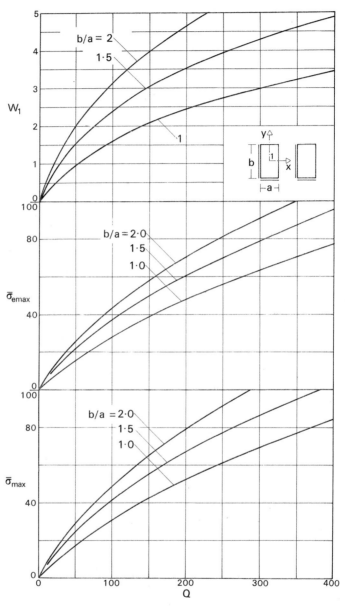

Fig. 4.40. Rectangular plates with unsymmetrical boundary conditions under uniform transverse pressure Q. Variations, with Q, of the transverse deflection W, maximum equivalent stress $\bar{\sigma}_{e\,max}$, and the maximum stress $\bar{\sigma}_{max}$.

Design Curves and Data 87

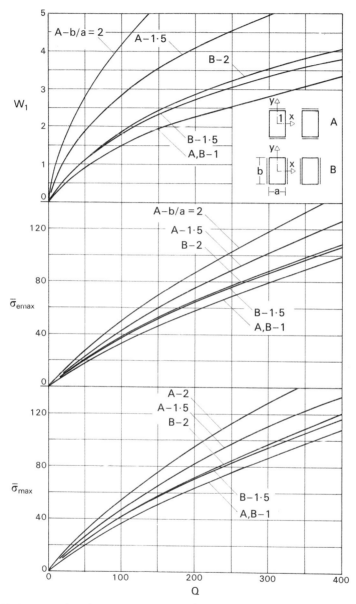

Fig. 4.41. Rectangular plates with unsymmetrical boundary conditions under uniform transverse pressure Q. Variations, with Q, of the transverse deflection W, maximum equivalent stress $\bar{\sigma}_{e\,max}$, and the maximum stress $\bar{\sigma}_{max}$.

88 Thin Plate Design For Transverse Loading

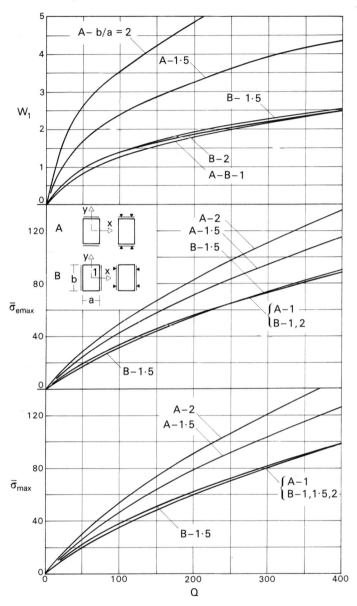

Fig. 4.42. Rectangular plates with unsymmetrical boundary conditions under uniform transverse pressure Q. Variations, with Q, of the transverse deflection W, maximum equivalent stress $\bar{\sigma}_{e\,max}$, and the maximum stress $\bar{\sigma}_{max}$.

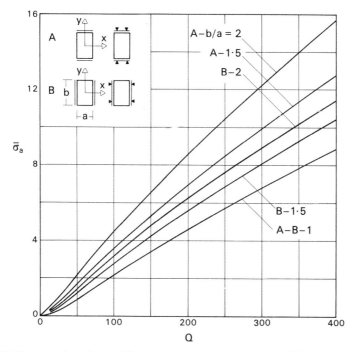

Fig. 4.43. Rectangular plates with unsymmetrical boundary conditions under uniform transverse loading Q. Variations, with Q, of the average reactive direct membrane stress $\bar{\sigma}_a$ developed at the boundaries.

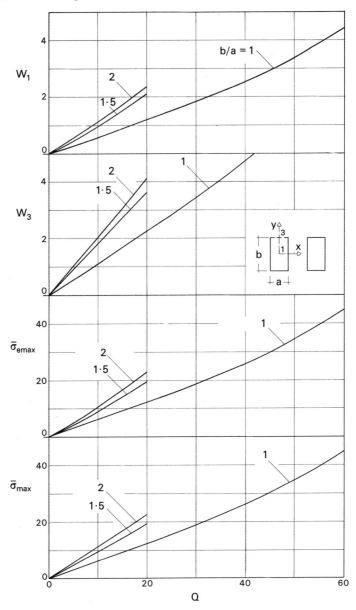

Fig. 4.44. Rectangular plates with unsymmetrical boundary conditions under uniform transverse pressure Q. Variations, with Q, of the transverse deflection W, maximum equivalent stress $\bar{\sigma}_{e\,max}$, and the maximum stress $\bar{\sigma}_{max}$.

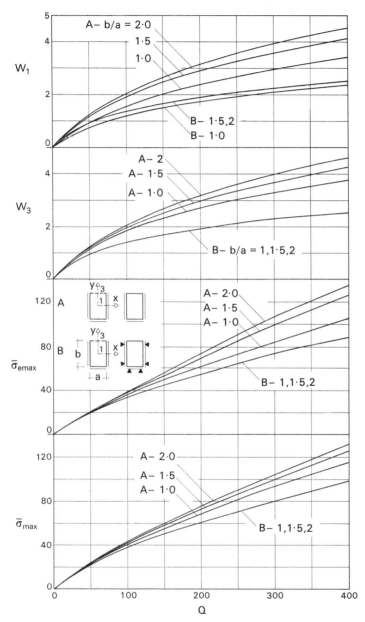

Fig. 4.45. Rectangular plates with unsymmetrical boundary conditions under uniform transverse pressure Q. Variations, with Q, of the transverse deflection W, maximum equivalent stress $\bar{\sigma}_{e\,\text{max}}$, and the maximum stress $\bar{\sigma}_{\text{max}}$.

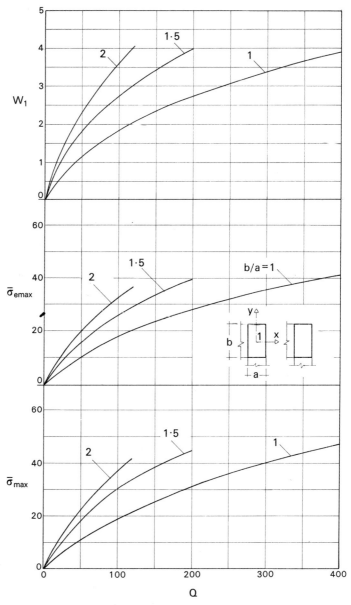

Fig. 4.46. Rectangular plates with unsymmetrical boundary conditions under uniform transverse pressure Q. Variations, with Q, of the transverse deflection W, maximum equivalent stress $\bar{\sigma}_{e\,max}$, and the maximum stress $\bar{\sigma}_{max}$.

Design Curves and Data 93

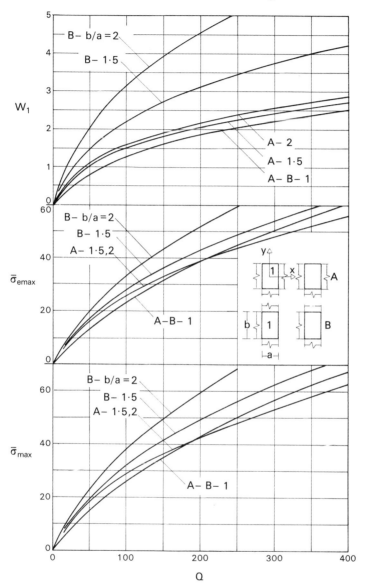

Fig. 4.47. Rectangular plates with unsymmetrical boundary conditions under uniform transverse pressure Q. Variations, with Q, of the transverse deflection W, maximum equivalent stress $\bar{\sigma}_{e\,\max}$, and the maximum stress $\bar{\sigma}_{\max}$.

4.4 Rectangular plates under a linearly varying distribution of transverse pressure (hydrostatic pressure)

4.4.1 Design curves

Figures 4.48–4.55 give design curves for rectangular plates under a linearly varying distribution of pressure, with zero intensity at the edge $y = b/2$ to maximum intensity Q_{max} at the edge $y = -b/2$. Aspect ratios considered are $b/a = 0.67, 1, 1.5, 2$. In using the design curves, it should be noted that the membrane boundary conditions of the two cases such as shown in Figs. 4.48 and 4.49 are in effect the same at the edges $y = -b/2$. This is due to the equilibrium of direct membrane forces in the y-direction ($\bar{\sigma}_{my}$) together with the condition of zero direct membrane stress at $y = b/2$.

4.4.2 Edge beams

For the case of the side walls being supported on edge beams at the top, an analysis is made by Aalami (1973) of the effects of the edge beam moment of inertia I on the stresses and deflections of the top edge. The analysis, which is conducted for square plates, concludes that the values of maximum equivalent stresses and the maximum tensile stresses are not significantly affected by the stiffness of the edge beam at the top, as these occur well away from the top edge.

The plate thickness may thus be evaluated on the basis of maximum stresses and the deflections at the plate centre with a fair accuracy without the use of the edge-beam moment of inertia. The top edge beam is subsequently to be designed to control the out-of-plane deflection of the plate at the top edge (W_3). Once the central deflection is evaluated, the moment of inertia of the top edge beam required to maintain the deflection at the mid-point of the top edge within a given ratio of central deflection may be determined, using Fig. 4.56 as a guide.

Figure 4.56 gives the top-edge deflection of a square plate with variations of the moment of inertia of the edge beam supporting that edge. The edge-beam moment of inertia I is expressed in terms of the plate thickness t and its side length a. The figure, being a guide to design, gives a band of deflection values bounded between the condition of thick plates (no large-deflection behaviour) and the condition

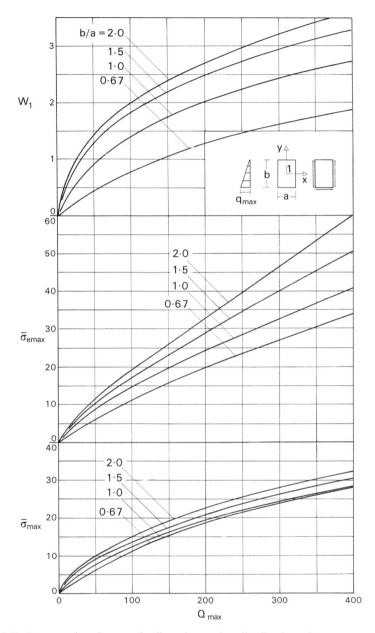

Fig. 4.48. Rectangular plates under linearly varying distribution of pressure (hydrostatic pressure). Variations with the transverse pressure Q_{max} of the out-of-plane deflections W, maximum equivalent stress $\bar{\sigma}_{e\,max}$ and the maximum tensile stress $\bar{\sigma}_{max}$.

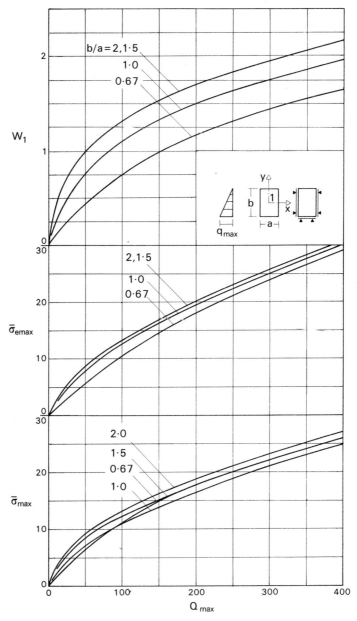

Fig. 4.49. Rectangular plates under linearly varying distribution of pressure (hydrostatic pressure). Variations with the transverse pressure Q_{max} of the out-of-plane deflections W, maximum equivalent stress $\bar{\sigma}_{emax}$ and the maximum tensile stress $\bar{\sigma}_{max}$.

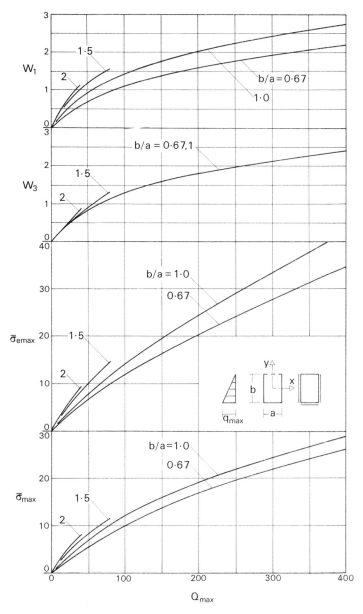

Fig. 4.50. Rectangular plates under linearly varying distribution of pressure (hydrostatic pressure). Variations with the transverse pressure Q_{max} of the out-of-plane deflections W, maximum equivalent stress $\bar{\sigma}_{e\,max}$ and the maximum tensile stress $\bar{\sigma}_{max}$.

98 Thin Plate Design For Transverse Loading

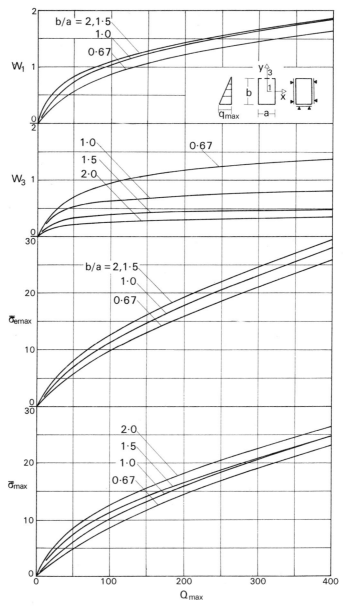

Fig. 4.51. Rectangular plates under linearly varying distribution of pressure (hydrostatic pressure). Variations with the transverse pressure Q_{max} of the out-of-plane deflections W, maximum equivalent stress $\bar{\sigma}_{e\,max}$ and the maximum tensile stress $\bar{\sigma}_{max}$.

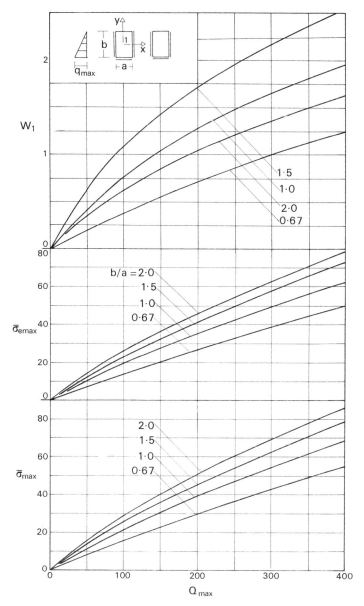

Fig. 4.52. Rectangular plates under linearly varying distribution of pressure (hydrostatic pressure). Variations with the transverse pressure Q_{max} of the out-of-plane deflections W, maximum equivalent stress $\bar{\sigma}_{e\,max}$ and the maximum tensile stress $\bar{\sigma}_{max}$.

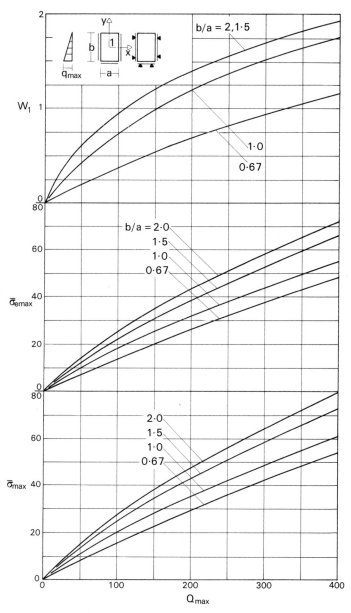

Fig. 4.53. Rectangular plates under linearly varying distribution of pressure (hydrostatic pressure). Variations with the transverse pressure Q_{max} of the out-of-plane deflections W, maximum equivalent stress $\bar{\sigma}_{e\,max}$ and the maximum tensile stress $\bar{\sigma}_{max}$.

Design Curves and Data 101

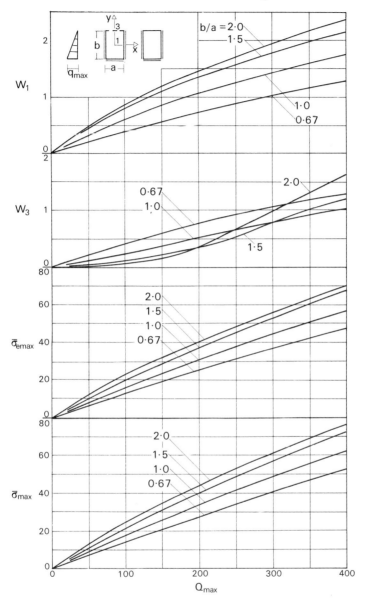

Fig. 4.54. Rectangular plates under linearly varying distribution of pressure (hydrostatic pressure). Variations with the transverse pressure Q_{max} of the out-of-plane deflections W, maximum equivalent stress $\bar{\sigma}_{e\,max}$ and the maximum tensile stress $\bar{\sigma}_{max}$.

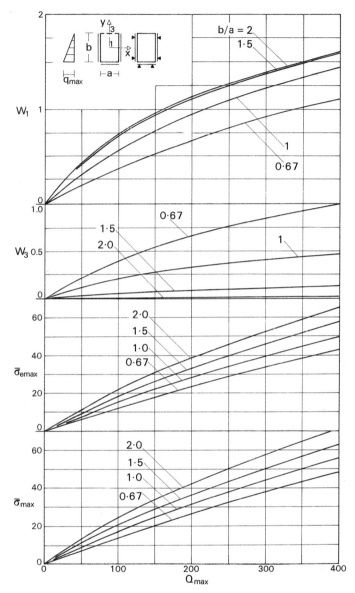

Fig. 4.55. Rectangular plates under linearly varying distribution of pressure (hydrostatic pressure). Variations with the transverse pressure Q_{max} of the out-of-plane deflections W, maximum equivalent stress $\bar{\sigma}_{e\,max}$ and the maximum tensile stress $\bar{\sigma}_{max}$.

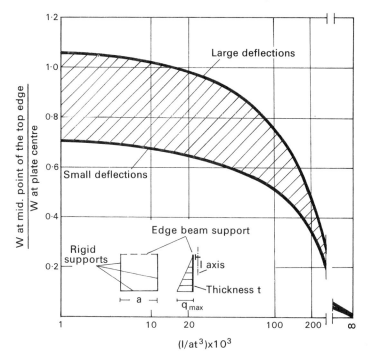

Fig. 4.56. Plates under linearly varying distribution of pressure. Variations, with the top edge beam moment of inertia I, of the out-of-plane deflection of the mid-point of the top edge.

of thin plates with extensive large-deflection behaviour (deflections several times plate thickness).

The reader will notice that for most practical dimensions the edge beams commonly provided are stiff enough to reduce the top-edge deflection to less than one half the deflection at plate centre.

After selecting a suitable I for the edge beam, it is next required to check the stresses developed in the edge beam. In the following, an approximate expression is given for the edge-beam moments. The expressions are derived on the assumption that the distribution of the plate reaction on the edge beam is as shown in Fig. 4.57 and given by the following expression:

$$V_x = 0{\cdot}5 V_{max}\left(1 + \cos\frac{2\pi x}{a}\right) \qquad (4.37)$$

104 Thin Plate Design For Transverse Loading

Further, if it is assumed that the edge beams are welded at their ends to the adjacent edge beams (zero rotation at the ends), from simple beam theory the moments developed are

At the corners (edge beam ends)

$$M = -28(EtI/a^2)W_3 \qquad (4.38)$$

At mid-span (mid-point of the top edge)

$$M = 17(EtI/a^2)W_3 \qquad (4.39)$$

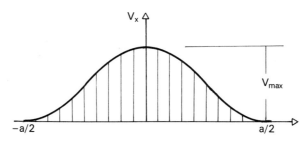

Fig. 4.57. Assumed distribution of reaction on the edge beam supporting the top edge of a plate under linearly varying distribution of pressure.

In the preceding two expressions W_3 is the deflection at the mid-point of the top edge, which may be obtained from Fig. 4.56, with the knowledge of central deflection and a trial value I.

4.4.3 Related work

The only work published in the literature and dealing explicitly with the large deflection of plates under hydrostatic pressure appears to be due to Davies (1969), who has treated the infinitely long plates by reducing the biaxial behaviour of the plate to a one-dimensional formulation. The related work from the present authors is reported by Aalami (1972c, 1973).

4.5 Circular plates under uniform transverse pressure or concentrated loadings

4.5.1 Design curves

For axisymmetrically loaded circular plates, four extreme boundary conditions are treated, two referring to simply supported plates

(Figs. 4.58, 4.59) and two to rotationally fixed conditions (Figs. 4.60, 4.61). In each case central deflections are given for a uniform distribution of pressure over the entire plate together with a concentrated loading at centre. Concentrated loadings considered are assumed to be uniformly distributed over an area with dimensions u/a equal to 0·05, 0·1, 0·15. The design curves presented are based on the approximate expressions given by Timoshenko and Woinowsky-Krieger (1959). In most cases the values of the maximum equivalent stress and the maximum tensile stress are the same from the approximate relationships used.

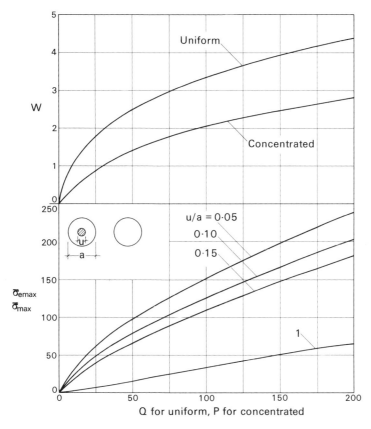

Fig. 4.58. Circular plates under transverse loading. Variations, with the transverse load, of central deflection W, maximum equivalent stress $\bar{\sigma}_{e\,max}$ and the maximum tensile stress $\bar{\sigma}_{max}$.

106 Thin Plate Design For Transverse Loading

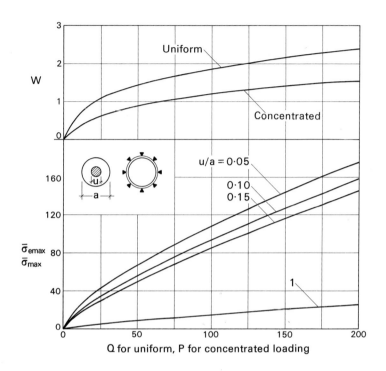

Fig. 4.59. Circular plates under transverse loading. Variations, with the transverse load, of central deflection W, maximum equivalent stress $\bar{\sigma}_{e\,max}$ and the maximum tensile stress $\bar{\sigma}_{max}$.

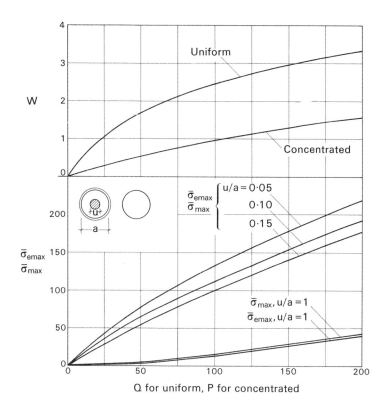

Fig. 4.60. Circular plates under transverse loading. Variations, with the transverse load, of central deflection W, maximum equivalent stress $\bar{\sigma}_{e\,\text{max}}$ and the maximum tensile stress $\bar{\sigma}_{\text{max}}$.

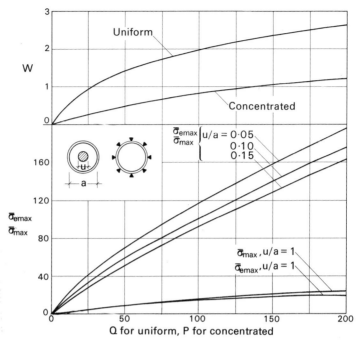

Fig. 4.61. Circular plates under transverse loading. Variations, with the transverse load, of central deflection W, maximum equivalent stress $\bar{\sigma}_{e\,max}$ and the maximum tensile stress $\bar{\sigma}_{max}$.

4.5.2 Related work

Much work has been done on the analysis and design of circular plates under transverse loading. For large-deflection elastic analysis, the reader may refer to Chien and Yeh (1954), Reissner (1947), Nash and Ho (1960), Schmidt (1968), Kao and Perrone (1971).

For the ultimate-strength elasto-plastic or rigid plastic behaviour of circular plates under transverse loading, the following literature may be consulted: Weil and Newmark (1955), Onat and Haythornthwaite (1956).

CHAPTER FIVE
Numerical Examples

The use and scope of design curves given in Chapter 4 are illustrated by the following numerical examples.

Example 5.1 Steel plates under uniform transverse pressure

Calculate the maximum deflection and the maximum equivalent stress of the following steel plate under uniform transverse pressure.

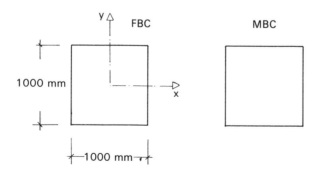

Fig. 5.1. Steel plate under uniform transverse pressure.

110 Thin Plate Design For Transverse Loading

Dimensions:

$a = 1000$ mm (39·4 in)
$b = 1000$ mm (39·4 in)
$t = 8$ mm (0·31 in)

Material:

High tensile steel, $E = 2·06 \times 10^5$ N/mm² (30 000 ksi), $\sigma_y = 448$ N/mm² (65 ksi), $v = 0·3$.

Boundary conditions (Fig. 5.1):

Flexural boundary conditions (FBC): All sides rigidly supported, rotationally free.
Membrane boundary conditions (MBC): Membrane direct and shearing stresses zero on all sides.

Loading:

Uniformly distributed pressure of intensity $q = 0·1$ N/mm² (14·5 psi).

Solution:

$b/a = 1000/1000 = 1$

$Q = a^4 q/t^4 E = (1000^4 \times 0·1)/(8^4 \times 2·06 \times 10^5) = 118$

From the list of problems treated in Fig. 3.1, the design quantities corresponding to this problem are given in Figs. 4.3, 4.6, 4.9.
For central deflection, from the upper section of Fig. 4.3,

$W_1 = 2·45$

$w_1 = W_1 t = 2·45 \times 8 = 19·6$ mm (0·772 in)

For the maximum equivalent stress, from the upper section of Fig. 4.6

$\bar{\sigma}_{e\,max} = 27$

$\sigma_{e\,max} = (t/a)^2 E \bar{\sigma}_{e\,max} = (8/1000)^2 \times 2·06 \times 10^5 \times 27$
$= 355$ N/mm² (51·5 ksi)

The location of this stress is, from the upper section of Fig. 4.15, near the plate corner. The equivalent stress at plate centre is, in this case, much lower than the maximum value near the corner. The reader interested in the evaluation of the equivalent stress at plate centre should consult numerical examples A11.1 and A11.2.

Example 5.2 Riveted aluminium plate under transverse loading

Consider a panel of aluminium alloy, with dimensions as shown in Fig. 5.2, riveted to stiffeners at the supports as shown in Fig. 2.5. Calculate the panel's central deflection, the maximum equivalent stress and the shearing forces in the rivets.

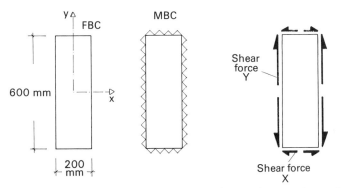

(i) Dimensions and boundary conditions   (ii) Reactive membrane shearing forces at the boundaries

Fig. 5.2. Aluminium plate under uniform transverse pressure.

Dimensions:

$a = 200$ mm (7·89 in)
$b = 600$ mm (23·6 in)
$t = 1·4$ mm (0·0551 in)

Material:

Aluminium alloy with $E = 8 \times 10^4$ N/mm^2 (11 600 ksi), proof stress 480 N/mm^2 (70 ksi), Poisson's ratio $= 0·316$ (assumed 0·3 in this example; for allowance for variations in Poisson's ratio see Example 5.4).

112 Thin Plate Design For Transverse Loading

Boundary conditions (Fig. 5.2):

FBC: Rigidly supported, rotationally free on all four edges.
MBC: Edges restrained against tangential movement by the rivets, zero direct membrane stress.

Loading:

Uniformly distributed air pressure equal to 0·04 N/mm² (5·8 psi).

Solution:

The loading and the boundary conditions of this example relate to the list included in Fig. 3.1, from which the corresponding design curves are given in Figs. 4.1, 4.4, 4.7, 4.11.

$b/a = 600/200 = 3$

$Q = a^4 q/t^4 E = (200^4 \times 0{\cdot}04)/(1{\cdot}4^4 \times 8 \times 10^4) = 208$

Central deflection: (middle section of Fig. 4.1)
The central deflection is greater than five times the plate thickness, it is therefore beyond the upper limit considered for deflections in the design curves. The value of deflection may be calculated from the supplementary design tables given in the Appendix.

From the last section of Table A3, and with the notation defined in the Appendix,

$W_1 = K_l K_r Q = 13{\cdot}42 \times 10^{-2} \times 0{\cdot}286 \times 208 = 7{\cdot}98$

Hence,

$w_1 = W_1 t = 7{\cdot}98 \times 1{\cdot}4 = 11{\cdot}2$ mm (0·44 in)

For the maximum equivalent stress, from Fig. 4.4

$\bar{\sigma}_{e\,max} = 99$

$\sigma_{e\,max} = (t/a)^2 E \bar{\sigma}_{e\,max} = (1{\cdot}4/200)^2 \times 8 \times 10^4 \times 99$

$= 388$ N/mm² (56·3 ksi)

For shearing forces in the rivets, from upper section of Fig. 4.11,

Average reactive membrane shear stress at $x = a/2$ is $\bar{\tau}_a = 5{\cdot}3$
Similarly shear stresses at $y = b/2$ are $\bar{\tau}_a = 8{\cdot}9$

Hence, shear force Y acting on half of the edge at $x = a/2$ is given by

$Y = (b/2)t\tau a = (b/2)t(t/a)^2 E\bar{\tau}_a$
$= (600/2) \times 1.4 \times (1.4/200)^2 \times 8 \times 10^4 \times 5.3 = 8726$ N
(1961 lbf)

which means that the rivets at the edge $x = a/2$ between $y = 0$ and $y = b/2$ should be designed to carry safely 8726 N (1961 lbf). Similarly, the design shear force at $y = b/2$ is given by

$X = (a/2)t(t/a)^2 E\bar{\tau}_a$
$= (200/2) \times 1.4 \times (1.4/200)^2 \times 8 \times 10^4 \times 8.9 = 4884$ N
(1097 lbf)

Example 5.3 Aspect ratios not specifically covered— corner panels

Calculate the central deflection, and the maximum equivalent stress of the following steel plate panel.

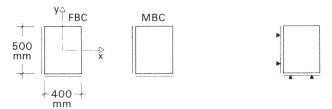

(i) Dimensions and boundary conditions   (ii) Alternative membrane boundary conditions

Fig. 5.3. Corner plate under uniform transverse pressure.

Dimensions:

$a = 400$ mm (15.7 in)
$b = 500$ mm (19.7 in)
$t = 4$ mm (0.157 in)

Material:

High tensile steel with $E = 2.06 \times 10^5$ N/mm² (30 000 ksi), $\sigma_y = 448$ N/mm² (65 ksi), Poisson's ratio $= 0.3$.

114   Thin Plate Design For Transverse Loading

Boundary conditions (Fig. 5.3):

FBC: The edges $x = -a/2$ and $y = -b/2$ on rigid supports, rotationally fixed. The edges $x = a/2$ and $y = b/2$ rigidly supported, rotationally free.
MBC: The edges $x = -a/2$ and $y = -b/2$ remain straight with the average membrane direct stress on them equal to zero, zero membrane shear stresses.

For the edges $x = a/2$ and $y = b/2$, zero membrane direct and shear stresses.

Note that the specified boundary conditions may be considered to approximate a corner panel from an array of panels on orthogonal supports. It should also be noted that, in this example, the MBC's are in effect the same as shown in Fig. 5.3(ii), in which the edges at $x = -a/2$ and $y = -b/2$ are held in position.

Loading:

Uniformly distributed loading of intensity $q = 0.13$ N/mm^2 (18.1 psi)

Solution:

The specifications of this problem relate from the directory of Fig. 3.3 to the design curves given in Fig. 4.40:

$b/a = 500/400 = 1.25$

$Q = a^4 q/t^4 E = 400^4 \times 0.13/4^4 \times 2.06 \times 10^5 = 63$

For central deflection (Fig. 4.40 upper section):

For aspect ratio 1, $W_1 = 1.20$
For aspect ratio 1.5, $W_1 = 1.80$
Using linear interpolation for aspect ratio 1.25

$$W_1 = 1.20 + \frac{1.25 - 1}{1.5 - 1}(1.80 - 1.20) = 1.50$$

$w_1 = W_1 t = 1.50 \times 4 = 6.0$ mm (0.236 in)

Similarly, for the maximum equivalent stress, from the middle of the same figure

For $b/a = 1$, $\sigma_{e\,max} = 19$

For $b/a = 1.5$, $\sigma_{e\,max} = 27$

$$\bar{\sigma}_{e\,max} = 19 + \frac{1\cdot 25 - 1}{1\cdot 5 - 1}(27 - 19) = 23$$

$$\sigma_{e\,max} = (t/a)^2 E \bar{\sigma}_{e\,max} = (4/400)^2 \times 2\cdot 06 \times 10^5 \times 23$$

$$= 474 \text{ N/mm}^2 \text{ (69 ksi)}$$

Example 5.4 Glass window of a high-rise building—variations in Poisson's ratio

With a safety factor of 2, recommend a thickness for a pane of glass under maximum wind loading of 1000 N/m² (0·145 psi) with the following specifications.

Dimensions:

$a = 1500$ mm (59·0 in)
$b = 2250$ mm (88·6 in)
$t =$ to be recommended

Material:

Glass with $E = 6\cdot 80 \times 10^4$ N/mm² (10 000 ksi), Poisson's ratio $= 0\cdot 22$, fracture tensile stress 50 N/mm² (7·25 ksi).

Boundary conditions (as shown in Fig. 2.4):

FBC: All edges rigidly supported, rotationally free.
MBC: Zero membrane direct and shear stresses on all edges.

Loading:

Uniformly distributed pressure $q = 1 \times 10^{-3}$ N/mm² (0·145 psi)

Solution:
The design is carried out for a Poisson's ratio of 0·3, subsequently the design quantities are checked for the correct Poisson's ratio.

$b/a = 2250/1500 = 1\cdot 5$

116 Thin Plate Design For Transverse Loading

For first trial, assume $t = 4$ mm (0·157 in)
Maximum permissible tensile stress

$$\sigma_{max} = \frac{\text{Fracture stress}}{\text{Factor of safety}} = \frac{50}{2} = 25 \text{ N/mm}^2 \text{ (3·62 ksi)}$$

$$\bar{\sigma}_{max} = (a/t)^2 \sigma_{max}/E = (1500/4)^2 \times 25/6·89 \times 10^4 = 51$$

Transverse pressure present

$$Q(\text{present}) = a^4 q/t^4 E = 1500^4 \times 1 \times 10^{-3}/4^4 \times 6·89 \times 10^4$$
$$= 287$$

From the upper section of Fig. 4.9, for $b/a = 1·5$, the value of the transverse pressure Q corresponding to the permissible $\sigma_{max} = 25$ N/mm^2 is given as

$$Q(\text{permissible}) = 210 < 287 \text{ present}$$

Hence the assumed thickness is not satisfactory.

For second trial, assume $t = 6$ mm (0·236 in)

$$\bar{\sigma}_{max}(\text{permissible}) = (1500/6)^2 \times 25/6·89 \times 10^4 = 22·6$$

$$Q(\text{present}) = 1500^4 \times 1 \times 10^{-3}/6^4 \times 6·89 \times 10^4 = 56·7$$

From Fig. 4.9

$$Q(\text{permissible}) = 80 > 56·7$$

Hence the thickness is satisfactory.
Adjustment for the correct value of Poisson's ratio $v^* = 0·22$
The central deflection, from the upper section of Fig. 4.3 and for $Q = 56·7$ is

$$W_1 = 2·6$$

The corrected central deflection for $v^* = 0·22$, using the relationship (3.11) is

$$W_1^* = 1·1(1 - 0·22^2) \times 2·6 = 2·72$$

$$w_1^* = 6 \times 2·72 = 16·3 \text{ mm (0·643 in)}$$

The actual value of the maximum tensile stress for $Q = 56{\cdot}7$ is, from Fig. 4.9,

$$\bar{\sigma}_{max} = 17$$

$$\sigma_{max} = (6/1500)^2 \times 6{\cdot}89 \times 10^4 \times 17 = 18{\cdot}7 \text{ N/mm}^2 \text{ (2}{\cdot}71 \text{ ksi)}$$

Now, for the adjusted maximum stress, using relationship (3.14)

$$\sigma^*_{max} = 1{\cdot}1(1 - 0{\cdot}3 \times 0{\cdot}22) \times 18{\cdot}7 = 19{\cdot}2 \text{ N/mm}^2 \text{ (2}{\cdot}87 \text{ ksi)}$$

In this case, as in most other practical cases, the slight variations in Poisson's ratio from the assumed value of 0·3 does not result in significant changes in design quantities.

Example 5.5 Steel plate under concentrated loading

Evaluate the central deflection and the maximum equivalent stress of a steel panel with dimensions and boundary conditions as shown in Fig. 5.4, under a central wheel loading of 14 000 N (3146 lbf).

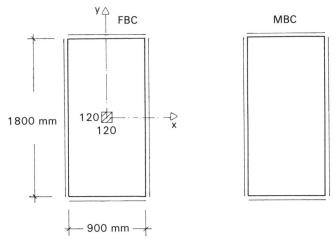

Fig. 5.4. Steel plate under central concentrated loading.

Dimensions:

$a = 900$ mm (35·4 in)
$b = 1800$ mm (70·9 in)
$t = 6$ mm (0·236 in)

118 Thin Plate Design For Transverse Loading

Material:

High tensile steel, $E = 2{\cdot}06 \times 10^5$ N/mm² (30 000 ksi), Poisson's ratio $= 0{\cdot}3$, yield stress $= 448$ N/mm² (65 ksi)

Boundary conditions:

As shown in Fig. 5.4.

Loading:

Concentrated loading assumed uniformly distributed over the area shown in the figure. $p = 14\,000$ N (3146 lbf)
$u = 120$ mm (4·72 in)
$v = 120$ mm (4·72 in)

Solution:

$P = a^2 p / t^4 E = 900^2 \times 14\,000/6^4 \times 2{\cdot}06 \times 10^5 = 42{\cdot}5$

$\alpha = u/a = 120/900 = 0{\cdot}133$

$\beta = v/a = 120/900 = 0{\cdot}133$

From the list of Fig. 3.2, the related design curves are to be found in Fig. 4.34.

For central deflection

For $\alpha \times \beta = 0{\cdot}1 \times 0{\cdot}1$, $W_1 = 1{\cdot}88$
For $\alpha \times \beta = 0{\cdot}2 \times 0{\cdot}2$, $W_1 = 1{\cdot}80$
Using linear interpolation for $\alpha \times \beta = 0{\cdot}133 \times 0{\cdot}133$
$W_1 = 1{\cdot}88 - \dfrac{0{\cdot}133 - 0{\cdot}1}{0{\cdot}2 - 0{\cdot}1} (1{\cdot}88 - 1{\cdot}80) = 1{\cdot}85$

$w_1 = 6 \times 1{\cdot}85 = 11{\cdot}1$ mm (0·437 in)

For maximum equivalent stress

For $\alpha \times \beta = 0{\cdot}1 \times 0{\cdot}1$, $\bar{\sigma}_{e\,\text{max}} = 48$
For $\alpha \times \beta = 0{\cdot}2 \times 0{\cdot}2$, $\bar{\sigma}_{e\,\text{max}} = 33$
For $\alpha \times \beta = 0{\cdot}133 \times 0{\cdot}133$

$\bar{\sigma}_{e\,\text{max}} = 48 - \dfrac{0{\cdot}133 - 0{\cdot}1}{0{\cdot}2 - 0{\cdot}1} (48 - 33) = 43{\cdot}5$

$\sigma_{e\,\text{max}} = (t/a)^2 E \bar{\sigma}_{e\,\text{max}} = (6/900)^2 \times 2{\cdot}06 \times 10^5 \times 43{\cdot}5$
$= 398$ N/mm² (57·7 ksi)

The reader is invited to compare the above solution with those obtained from the boundary conditions of the plate being continuous as given in Fig. 4.36.

Example 5.6 Side wall of a liquid container resting on an edge beam at top

Check the maximum equivalent stress in the side wall (Fig. 5.5) of the liquid container, and find the required moment of inertia of the edge beam to reduce the deflection of the top edge to a value about 40% of the corresponding deflection at plate centre.

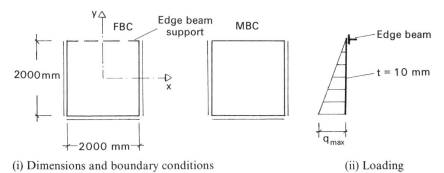

(i) Dimensions and boundary conditions (ii) Loading

Fig. 5.5. Side wall of a liquid container supported on an edge beam at the top.

Dimensions:

$a = 2000$ mm (78·8 in)
$b = 2000$ mm (78·8 in)
$t = 10$ mm (0·394 in)

Material:

High tensile steel, $E = 2·06 \times 10^5$ N/mm² (30 000 ksi), Poisson's ratio $= 0·3$, yield stress $= 248$ N/mm² (36 ksi)

Boundary conditions:

As shown in Fig. 5.5 symbolically.

Loading:

Water pressure with the maximum intensity q_{max} at the bottom edge
$q_{max} = 9 \cdot 805 \times 10^{-6} \times 2000 = 0 \cdot 0196$ N/mm² (2·84 psi)

Solution:

$b/a = 2000/2000 = 1$

$Q_{max} = a^4 q_{max}/t^4 E = 2000^4 \times 0 \cdot 0196/10^4 \times 2 \cdot 06 \times 10^5 = 152$

The problem under consideration is a case between the two extreme boundary conditions of the top edge being free (design curves of Fig. 4.54) and the extreme condition of the top edge resting on a rigid support (design curves of Fig. 4.52).

For central deflection

For top edge on a rigid support, $W_1 = 1 \cdot 1$
For top edge unsupported (free), $W_1 = 0 \cdot 85$
The design central deflection is approximated as the average of the preceding two extremes; thus

$W_1 = 0 \cdot 5(1 \cdot 1 + 0 \cdot 85) = 0 \cdot 975$

For the maximum equivalent stress

For top edge on a rigid support, $\bar{\sigma}_{e\,max} = 28$
For top edge unsupported (free), $\bar{\sigma}_{e\,max} = 24$
The design stress is approximated as

$\bar{\sigma}_{e\,max} = 0 \cdot 5(28 + 24) = 26$; hence

$\sigma_{e\,max} = (10/2000)^2 \times 2 \cdot 06 \times 10^5 \times 26 = 133 \cdot 9$ N/mm²
(19·4 ksi)

Edge beam design
It is required to maintain the following condition

$$\frac{W \text{ at midpoint of the top edge}}{W \text{ at plate centre}} = 0 \cdot 4$$

From Fig. 4.56, the coefficient for the required edge beam moment of inertia I may be approximated as 200, between the values relating to the small-deflections limit and the large-deflections extreme, hence

$(I/at^3) \times 10^3 = 200$

$I = \dfrac{200 \times 2000 \times 10^3}{1000} = 4 \times 10^5 \text{ mm}^4 \ (0.96 \text{ in}^4)$

Example 5.7 Penetration of yielding through the thickness

What is the maximum uniformly distributed pressure the square steel plate of Fig. 5.6 can sustain if local yielding is allowed to penetrate to a depth of one quarter thickness?

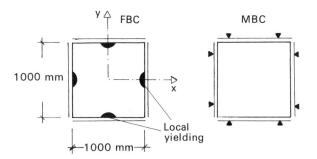

Fig. 5.6. Clamped steel plate under uniform transverse pressure.

Dimensions:

$a = 1000$ mm (39·4 in)
$b = 1000$ mm (39·4 in)
$t = 16$ mm (0·63 in)

Material:

Mild steel with $E = 2.06 \times 10^5$ N/mm² (30 000 ksi), Poisson's ratio $= 0.3$, yield stress 248 N/mm² (36 ksi).

Loading:

Uniformly distributed pressure q. Maximum intensity to be evaluated.

122 Thin Plate Design For Transverse Loading

Boundary conditions:

FBC: All edges rigidly supported, rotationally fixed.
MBC: Edges held in position, with the membrane shear stress equal to zero.

Solution:

Depth of penetration allowed for local yielding $\beta = 0.25$, hence the coefficient of local overstress α is given from the relationship (2.3).

$\alpha = 1 + 2\beta - 2\beta^2 = 1 + 2 \times 0.25 - 2 \times 0.25^2 = 1.375$

Maximum equivalent stress to be considered for design

$\sigma_{e\,max} = 1.375 \times 248 = 341 \text{ N/mm}^2$ (49.5 ksi)

$\bar{\sigma}_{e\,max} = (a/t)^2 \sigma_{e\,max}/E = (1000/16)^2 \times 341/2.06 \times 10^5 = 6.46$

From the design curves at the bottom section of Fig. 4.5

$Q = 23$; hence

$q = t^4 EQ/a^4 = 16^4 \times 2.06 \times 10^5 \times 23/1000^4 = 0.310 \text{ N/mm}^2$
(45 psi)

References and Bibliography

Aalami, B. and Chapman, J. C. (1969a) 'Large Deflection Behaviour of Orthotropic Plates Under Inplane and Transverse Loading'. *Proceedings of the Institution of Civil Engineers*, Vol. 42, March, pp. 347–382. Discussion in Vol. 44, November 1969, pp. 263–264.

Aalami, B. and Chapman, J. C. (1969b) 'Large Deflection Behaviour of Ship Plates Under Normal Pressure and Inplane Loading', with discussions. *Transactions of the Royal Institution of Naval Architects*, Vol. 114, April, pp. 151–181.

Aalami, B. (1972a) 'Large Deflection of Elastic Plates Under Patch Loading', *Journal of the Structural Division*, A.S.C.E., Vol. 98, No. ST 11, November, pp. 2567–2586.

Aalami, B., Mokhtarzadeh, A. and Mahmudi-Saati, P. (1972b) 'On Strength Design of Ship Plates Under Inplane and Transverse Loading'. *Transactions of the Royal Institution of Naval Architects*, Supplementary Papers, Vol. 114, November, pp. 519–534.

Aalami, B. (1972c) 'Large Deflection of Plates Under Hydrostatic Pressure—with Special Reference to Liquid Containers'. *Journal of Ship Research*, U.S.A., No. 4, Vol. 16, December, pp. 261–270.

Aalami, B. (1973) 'Analysis and Design of Cubic Liquid Containers'.

Conference on Stress and Strain in Engineering, Brisbane, *Institution of Engineers, Australia, National Conference Publication*, No. 73/5, pp. 77–82.

Barbanov, N. (undated) *Structural Design of Sea-Going Ships*. Mir Publishers, Moscow, p. 463.

Bares, R. (1971) *Tables for the Analysis of Plates, Slabs and Diaphragms Based on the Elastic Theory*. Bauverlag GmbH, Wiesbaden.

Basu, A. K. and Chapman, J. C. (1936) 'Large Deflection Behaviour of Transversely Loaded Rectangular Orthotropic Plates'. *Proceedings of the Institution of Civil Engineers*, Vol. 35, September, pp. 79–110.

Bauer, F., Bauer, L., Becker, W. and Reiss, E. L. (1964) 'Bending of Rectangular Plates with Finite Deflections'. *New York University Courant Institute of Mathematical Sciences*, IMM-NYU 322, April, pp. 1–32.

Berger, H. M. (1955) 'A New Approach to the Analysis of Large Deflections of Plates'. *Journal of Applied Mechanics*, December, pp. 465–472.

Bunting, S., Briscoe, P. H., Subbiah, Lt. J. and Glenton, S. J. (1972) 'Flat Plates Subject to Localised Loads'. *Royal Institution of Naval Architects*. Supplementary Paper, Vol. 114, pp. 499–506.

Burgoyne, C. J. (1972) *An Investigation into the Behaviour of Elasto-Plastic Plates*. A thesis submitted to the Imperial College of Science and Technology in partial fulfillment of the requirements for the Master of Science Degree, Structural Engineering Laboratories, Imperial College, University of London, August, pp. 1–61.

Chia, C. Y. (1971) 'Nonlinear Theory of Heterogeneous and Anisotropic Plates'. *Proceedings of the Third Australian Conference on the Mechanics of Structures and Materials*, Vol. II, Session B2—Plate Theory, University of Auckland, August, pp. 1–25.

Chia, C. Y. (1972a) 'Large Deflection of Rectangular Orthotropic Plates'. *Journal of the Engineering Mechanics Division*. Proceedings of the American Society of Civil Engineers, Vol. 98, No. EM5, October, pp. 1286–1298.

Chia, C. Y. (1972b) 'Finite Deflections of Uniformly Loaded, Clamped, Rectangular, Anisotropic Plates'. *AIAA Journal*, Vol. 10, No. 11, November, pp. 1399–1400.

Chien, W. Z. (1955) *Problem of Large Deflection of Circular Plate*.

Presented before the Third Scientific Congress organised at Karpacz, August, pp. 3–11.

Chien, W. Z. and Yeh, K. Y. (1954) 'On the Large Deflection of Circular Plate'. *Scientia Sinica*, Vol. III, pp. 405–436.

Clarkson, J. (1956) 'A New Approach to the Design of Plates to Withstand Lateral Pressure'. *Trans. INA*, Vol. 98, pp. 443–463.

Clarkson, J. (1963) 'Tests of Flat Plated Grillages Under Uniform Pressure'. *Meeting of the Royal Institution of Naval Architects*, Paper No. 8, March, pp. 1–14.

Crisfield, M. A. (1973) 'Large-Deflection Elasto-Plastic Buckling Analysis of Plates Using Finite Elements'. *Transport and Road Research Laboratory*, Department of the Environment. T.R.R.L. Report LR 593, pp. 1–50.

Davies, J. D. (1969) 'Large Deflexion Analysis of Plates with Particular Application to Long Rectangular Containers'. *International Symposium on Thin-Walled Steel Structures*, Crosby Lockwood, London.

Eggwertz, S. and Norr, A. (1953) 'Analysis of Thin Square Plates Under Normal Pressure and Provided with Edge Frames of Finite Stiffnesses in the Plane of the Plates'. *The Aeronautical Research Institute of Sweden*, Report No. 50, pp. 1–34.

Feodosyev, V. (1964) *Strength of Materials*. Mir Publishers, Moscow, pp. 1–570.

Fluegge, W. (1962) *Handbook of Engineering Mechanics*. McGraw-Hill Book Company, London.

Frownfelter, C. R. (1959) 'Structural Testing of Large Glass Installations'. *ASTM Special Technical Publication*, No. 251, pp. 19–30.

Girkmann, K. (1963) *Flaechentragwerke*. Springer-Verlag, Vienna.

Illyushin, A. A. (1956) *Plasticite*. Eyrolles, Paris, pp. 170–180.

Jaeger, L. G. (1958) 'An approximate Analysis for Plating Panels Under Uniformly Distributed Load'. Paper No. 6254, pp. 137–144.

Jones, N. and Walters, R. M. (1971) 'Large Deflections of Rectangular Plates'. *Journal of Ship Research*, June, pp. 164–171.

Kaiser, R. (1936) 'Research and Experimental Study on Stresses and Deformation of Square Plates'. *Zeitung für Angewandte Mathematik und Mechanik*, Vol. 16, April, pp. 73–98.

Kao, R. and Perrone, N. (1971) 'Large Deflections of Axisymmetric Circular Membranes'. *International Journal of Solids Structures*, Vol. 7, pp. 1601–1612.

Levy, S. (1942) 'Bending of Rectangular Plates with Large Deflections'. *NACA Report* 737.

Lin, T. H., Lin, S. R. and Mazelsky, B. (1972) 'Elastoplastic Bending of Rectangular Plates with Large Deflection'. *Journal of Applied Mechanics*, December, pp. 978–982.

Liptak, T., Ballantyne, E. R. and Brotchie, J. F. (1973) 'Combined Membrane and Bending Theory Applied to Rectangles of Window Glass Subjected to Wind Pressures'. *National Conference Publication* No. 73/5, Conference on Stress and Strain in Engineering, Brisbane, Institution of Engineers, pp. 83–88.

Mallet, R. H. and Marcal, P. V. (1968) 'Finite Element Analysis of Non-Linear Structures'. *Journal of the Structural Division*, A.S.C.E., Vol. 94, No. ST9, Proc. Paper 6115, September, pp. 2081–2105.

Massonnet, C. H. (1968) 'General Theory of Elasto-Plastic Membrane Plates'. In *Engineering Plasticity*, ed. Heyman, J. and Leckie, F. A. Cambridge University Press, pp. 443–471.

Merrison, A. W. *et al.* (1973) *Inquiry into the Design and Erection of Steel Box Girder Bridges*, H.M.S.O., London.

Murray, D. W. and Wilson, E. L. (1969) 'Finite Element Large Deflection Analysis of Plates'. *Journal of the Engineering Mechanics Division*, A.S.C.E., Vol. 95, No. EM1, Proc. Paper 6398, February, pp. 143–165.

Nash, W. A. and Ho, F. H. (1960) 'Finite Deflections of a Clamped Circular Plate on an Elastic Foundation'. *VI Congress International Association of Bridge and Structural Engineers Final Report*, pp. 61–71.

Niyogi, A. K. (1973) 'Nonlinear Bending of Rectangular Orthotropic Plates'. *International Journal of Solids Structures*, Vol. 9, pp. 1133–1139.

Onat, E. T. and Haythornthwaite, R. M. (1956) 'The Load-Carrying Capacity of Circular Plates at Large Deflection'. *Journal of Applied Mechanics*, March, pp. 49–55.

Otter, J. R. H. (1965) 'Computations for Prestressed Concrete Reactor Pressure Vessels Using Dynamic Relaxation'. *Nuclear Structural Engineering*, Amsterdam, Vol. 1, No. 1, pp. 143–165.

Otter, J. R. H. and Cassell, A. L. (1966) 'Dynamic Relaxation'. *Proceedings of the Institution of Civil Engineers*, Vol. 35, December, pp. 633–656.

PPG Industries (undated) *Glass Product Recommendations*. P.P.G. Industries Technical Service Report No. 101—Structural, pp. 1–27.

Reissner, E. (1947) 'On Finite Deflections of Circular Plates'. Proceedings of the First Symposium in Applied Mathematics of the American Mathematical Society, Brown University, Providence, R.I., August, pp. 213–219.

Rushton, K. R. (1969a) 'Dynamic Relaxation Solution for the Large Deflection of Plates with Specified Boundary Stresses'. *Journal of Strain Analysis*, Vol. 4, No. 2, pp. 25–32.

Rushton, K. R. (1969b) 'Dynamic-Relaxation Solutions of Elastic-Plate Problems'. *Journal of Strain Analysis*, Vol. 3, No. 1, pp. 23–32.

Schmidt, R. (1968) 'Large Deflections of a Clamped Circular Plate'. *Engineering Mechanics Division, Proceeding of the American Society of Civil Engineers*, Vol. 94, No. EM6, December, pp. 1603–1606.

Scholes, A. and Bernstein, E. L. (1969) 'Bending of Normally Loaded Simply Supported Rectangular Plates in the Large-Deflection Range'. *Journal of Strain Analysis*, Vol. 4, No. 3, pp. 190–198.

Smith, W. (1963) 'Investigations into the Use of Fork Lift Trucks on Board Ship'. *Lloyd's Register of Shipping*, Paper No. 6, pp. 1–17.

Steinhardt, O. and Abdel-Sayed, G. (1963) *Zur Tragfaehigkeit von versteiften Flachblechtafeln im Metalbau*, Germany, Vesuchsanstalt fuer Stahl, Holz und Steine der TH Fredericiana, Karlsruhe, p. 147.

Timoshenko, S. P. and Woinowsky-Krieger, S. (1959) *Theory of Plates and Shells*. McGraw-Hill Book Company, London.

Von Karman, T. (1910) 'Festigkeitsprobleme im Maschinenbau'. *Encyklopaedie der Mathematischen Wissenschaften*, Vol. 4, p. 349.

Wah, T. (1958) 'Large Deflection Theory of Elasto-Plastic Plates'. *Journal of the Engineering Mechanics Division*, Paper 1822. Proceedings of the American Society of Civil Engineers, Vol. 84, No. EM 4, October, pp. 1–24.

Wah, T. (1960) 'A Theory for the Plastic Design of Ship Plating Under Uniform Pressure'. *Journal of Ship Research*, November, pp. 17–24.

Wang, C. T. (1948) 'Non-Linear Large Deflection Boundary Value Problems of Rectangular Plates'. NACA TN 1425.

Way, S. (1933) 'Bending of Circular Plates with Large Deflection'. *Annual Meeting of the American Society of Mechanical Engineers*, December, pp. 627–636.

Way, S. (1938) 'Uniformly Loaded Clamped Rectangular Plates with Large Deflections'. *Proceedings of the Fifth International Congress of Applied Mechanics*, Cambridge, Mass., pp. 123–128.

Weil, N. A. and Newmark, N. M. (1955) 'Large Plastic Deformations of Circular Membranes'. *Journal of Applied Mechanics*, December, pp. 533–538.

Weiss, S. (1969) 'Gleichmaessig innerhalb eines Rechtecks symmetrisch belastete, elastisch eingespannte Platte bei grossen Durchbiegungen'. *Schiffstechnik*, Band 16, Heft 82, pp. 59–70.

Williams, D. G. (1971) 'Some Examples of the Elastic Behaviour of Initially Deformed Bridge Panels'. *Structural Engineering Division Report*, Imperial College of Science and Technology, London.

Williams, M. L. (1955) 'Large Deflection Analysis for a Plate Strip Subjected to Normal Pressure and Heating'. *Journal of Applied Mechanics*, December, pp. 458–464.

Zienkiewicz, O. C. (1971) *The Finite Element Method in Engineering Science*. McGraw-Hill Book Company, London, p. 340.

APPENDIX
Data Tables for Large-deflection Elastic Behaviour of Transversely Loaded Plates

A1 Notation

a	plate side in x-direction
b	plate side in y-direction
E	Young's modulus of elasticity
FBC	flexural boundary conditions
K_l	linear coefficients from small deflection solutions
K_r	large deflection reduction coefficients
MBC	membrane boundary conditions
P	$a^2 p / t^4 E$
p	total magnitude of applied patch loading
Q	$a^4 q / t^4 E$
q	intensity of transverse loading
t	plate thickness
UDL	uniformly distributed loading
u	length of patch loading in x-direction
v	length of patch loading in y-direction
W	w/t
w	transverse deflection of plate
x, y	rectangular coordinates
α	u/a

β	v/a
v	Poisson's ratio
σ_x, σ_y	stresses in x- and y-directions
$\bar{\sigma}_x$	$(a/t)^2 \sigma_x / E$
$\bar{\sigma}_y$	$(a/t)^2 \sigma_y / E$

Subscripts and superscripts

1, 2, 3,...	refer to locations on plate
b	refers to bending action
e	equivalent stress
m	refers to membrane action
x, y	coordinate directions
*	quantities referring to Poisson's ratio v^*

A2 General description

A comprehensive range of data is presented in tabular form for the elastic large-deflection behaviour of rectangular plates under transverse loading. These data greatly facilitate the analysis and design of plated structures under transverse loading. They are applicable to metal plates and also to isotropic plates composed of other materials, such as plywood, glass or plastics. The tables may be used directly to evaluate bending and membrane stresses in the analysis of plate problems, or they may be used to supplement the graphical-design procedure discussed in the previous chapters. Researchers in the general field of plate behaviour will find the tables useful as a basis for comparison with particular solutions. For each plate problem considered, the tables list the values of central deflection, deflection at mid-point of the free edge (where applicable), bending and membrane stresses at centre for increasing values of transverse pressure. When they are of design significance, boundary stresses are also tabulated. Maximum stresses together with equivalent stress at plate centre, and in most cases at another selected point, may be evaluated directly from the results given. The maximum equivalent stress and the maximum tensile stress anywhere in the plate are given graphically in the previous sections. However, in some cases, the maximum direct stress and/or equivalent stress may occur at the plate centre or the selected point on the boundary for which tabular results for the stress components are given.

In the following sections, the data tables are described in detail and several numerical examples are given to illustrate their use.

A3 Boundary conditions

The boundary conditions selected are the same as those discussed in the main text, for which a simple design procedure is outlined already. These boundary conditions are considered to be either of direct practical importance or to provide an upper or lower limit to the more common plate problems. The boundary conditions relating to each data table are represented diagrammatically at the top left corner of each table, the left-hand diagram giving the flexural boundary conditions (FBC), and the right-hand diagram giving the membrane boundary conditions (MBC). For a rapid reference, the symbols used in the latter diagrams are summarised in Fig. A1. The cases for

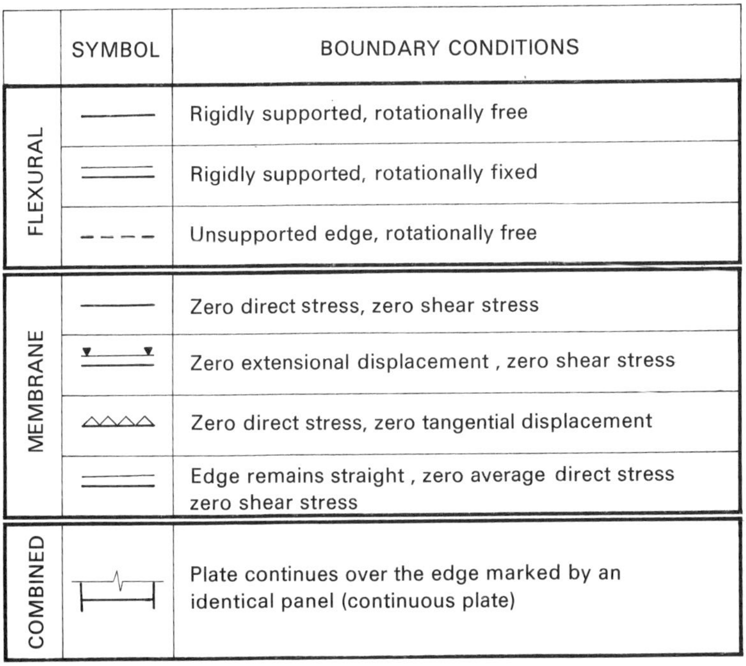

	SYMBOL	BOUNDARY CONDITIONS
FLEXURAL	———	Rigidly supported, rotationally free
	═══	Rigidly supported, rotationally fixed
	----	Unsupported edge, rotationally free
MEMBRANE	———	Zero direct stress, zero shear stress
	▼——▼	Zero extensional displacement, zero shear stress
	△△△△	Zero direct stress, zero tangential displacement
	═══	Edge remains straight, zero average direct stress, zero shear stress
COMBINED	⊢⋀⊣	Plate continues over the edge marked by an identical panel (continuous plate)

Fig. A1. Legend of basic symbols used for representation of flexural and membrane boundary conditions.

132 Thin Plate Design For Transverse Loading

which data are presented, together with the appropriate table number, are summarised in Figs. A2(i–v). For a specific plate problem, the user must define the boundary conditions, the type of applied loading and the aspect ratio; then the number of the corresponding data table may be read directly from Figs. A2(i–v). For example, for a rectangular plate $b/a = 2$ with FBC of rigid supports, rotationally fixed, and MBC of zero membrane shear stress and zero in-plane edge displacement, under a central patch loading of $\alpha \times \beta = 0.3 \times 0.3$, Fig. A2(ii) locates the required data set in Table A16.

A4 Loading

The data given in this volume are for plates acted upon by transverse loading, i.e., loading perpendicular to the plane of the plate. In the case of some membrane boundary conditions, in-plane forces develop at the boundaries which are reacted by the surrounding structure.

Plate problems with symmetrical boundary conditions and uniformly distributed loading (UDL) on the entire plate are listed in Fig. A2(i). Figure A2(ii) contains the cases of plates under a central patch loading distributed uniformly over the patch area. Figure

Boundary conditions		b/a	Boundary conditions		b/a
Flexural	Membrane	1, 1·5, 2, 3	Flexural	Membrane	1, 1·5, 2, 3
		A1			A5
		A2			A6
		A3			A7
		A4			A8
Uniformly distributed pressure					A9

Fig. A2(i). Number of data tables for plates with symmetrical boundary conditions under uniform transverse pressure.

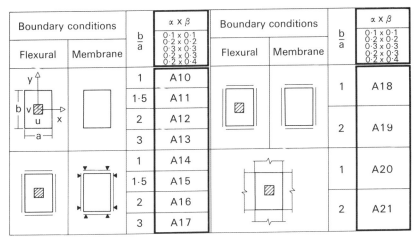

Fig. A2(ii). Number of data tables for plates with symmetrical boundary conditions under central patch loading.

Fig. A2(iii). Number of data tables for plates with unsymmetrical boundary conditions under uniform transverse pressure.

134 Thin Plate Design For Transverse Loading

Boundary conditions		b/a	Boundary conditions		b/a
Flexural	Membrane	0·67,1,1·5,2	Flexural	Membrane	0·67,1,1·5,2
↑b ↦a↤	☐	A35	[ ]	☐	A39
	▶☐◀ ▲▼	A36		▶☐◀ ▲▼	A40
[ ]	☐	A37	[ ]	☐	A41
	▶☐◀ ▲▼	A38		▶☐◀ ▲▼	A42

Fig. A2(iv). Number of data tables for plates under linearly varying distribution of pressure (zero at top and maximum at bottom).

A2(iii) is for rectangular plates with unsymmetrical boundary conditions. Plates loaded by a linearly varying distribution of transverse pressure are given in Fig. A2(iv).

For each plate problem, large-deflection data are given for increasing values of transverse pressure Q or transverse loading P. The range of the non-dimensional transverse loadings considered cover most structural applications. The upper limit in any particular case is usually determined from consideration of loading restrictions, maximum allowable stress or transverse deflection.

Deflections and stresses for transverse pressures between the values quoted in the tables may be approximated by linear interpolation.

A5   Small-deflection data—linear coefficients K_l

For each plate problem, the particular large-deflection stress or deflection is related to a corresponding small-deflection stress or deflection coefficient, which is called the linear coefficient K_l. Values of K_l are given on the top row of each large-deflection data set. Small-deflection solutions do not account for membrane action, and therefore they apply only to flexural behaviour (deflections and bending stresses). For the cases considered herein, small-deflection membrane stresses are all equal to zero. It should be noted that the K_l coefficient

Appendix 135

given for membrane stresses relates only to large-deflection behaviour, as will be discussed later, and should be disregarded when analysing small-deflection behaviour.

For small-deflection behaviour the following relationships apply:

For deflections

$$W = K_l Q \qquad (A5.1)$$

hence

$$w = K_l Q t = K_l (q a^4 / E t^3) \qquad (A5.2)$$

For bending stresses

$$\bar{\sigma}_b = K_l Q \qquad (A5.3)$$

hence

$$\sigma_b = K_l (t/a)^2 E Q = K_l (a/t)^2 q \qquad (A5.4)$$

For membrane stresses

$$\sigma_m = 0 \qquad (A5.5)$$

As an example, consider the case of a square ($b/a = 1$) simply supported plate under uniform transverse pressure (FBC = rigid supports, rotationally free; MBC = zero membrane direct and shear stresses). The problem relates to the first case of Table A1, from which

For central deflection $K_e = 4 \cdot 434 \times 10^{-2}$; hence

$$w_1 = 4 \cdot 434 \times 10^{-2} (q a^4 / E t^3)$$

The corresponding coefficient from Timoshenko and Woinowsky-Krieger (1959) (p. 120, Table 8) is

$$12(1 - 0 \cdot 3^2) \times 0 \cdot 004\ 06 = 4 \cdot 433 \times 10^{-2}$$

For bending stresses at centre

$$\sigma_{bx1} = \sigma_{by1} = 0 \cdot 286 (a/t)^2 q$$

The corresponding coefficient from the latter text is

$$6 \times 0 \cdot 0479 = 0 \cdot 2874$$

A6 Large-deflection data—reduction coefficients K_r

Non-dimensional large-deflection coefficients are presented in a specially devised form, called reduction coefficients, such that their value may be interpreted to express the extent of the development of membrane forces, and the resulting deviation of the plate behaviour from classical linear small-deflection behaviour. Prior to the explanation of the use of the large-deflection reduction coefficients, the following brief description of their evaluation will give a better appreciation of their significance. Consider a typical variation of a large-deflection solution for bending stress or deflection with the transverse pressure Q as shown in Fig. A3. For any given value of Q, the ratio of the

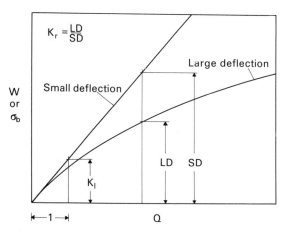

Fig. A3. Pictorial illustration of large-deflection reduction coefficients K_r and small-deflection (linear) coefficients K_l.

ordinate obtained from a large-deflection solution (designated LD) to a corresponding small-deflection solution (designated SD) signifies the magnitude of deviation of large deflection behaviour from its corresponding small-deflection value, and is called herein the large deflection reduction coefficient K_r. Hence

$$K_r = \frac{\text{large-deflection value}}{\text{small-deflection value}} \qquad (A6.1)$$

Clearly, K_r varies with transverse pressure Q. Furthermore, for any given value of Q, K_r's for deflection and stresses at one point are, in general, different from the K_r's for deflection and stresses at another point of the same plate. As an example, for the square simply supported plate of Table A1 under uniform transverse loading with intensity of $Q = 300$, K_r for central deflection is 0·304, which signifies that at this loading central deflection is 30·4% of its corresponding small-deflection value. Similarly, central bending stresses from the same row of Table A1 are 16·8% of the values given by small deflection theory.

Now, for membrane stresses it is necessary to know how they compare with the bending stresses for the same loading and if possible at the same point. The membrane stresses are, therefore, expressed as a ratio of the small-deflection bending stresses at the same point. For example, from Table A1 for $Q = 300$, the large-deflection reduction coefficient for $\bar{\sigma}_{mx1}$ is $K_r = 0\cdot117$, which means that the magnitude of the central membrane stress is 11·7% of the small-deflection bending stress at the same point. The large-deflection reduction coefficients 0·117 for $\bar{\sigma}_{mx1}$ and 0·168 for $\bar{\sigma}_{bx1}$ may, therefore, be compared directly with one another to give the ratio of the membrane to bending stresses at this point. For this example, the ratio of the large-deflection membrane stress developed to the existing bending stress is 0·117/0·168 = 0·696, about 70%.

At points where membrane stresses develop, but bending stresses are zero (such as $\bar{\sigma}_{my2}$ for cases considered in Table A1) the membrane stresses may no longer be related to the bending stresses at the same point. In these cases the large-deflection reduction coefficients are expressed such as to relate the magnitude of the membrane stress to the bending stress at plate centre in the same direction and at the same loading. For example, for a square simply supported plate under a transverse pressure of $Q = 300$, Table A1 gives $K_r = -0\cdot335$ for $\bar{\sigma}_{my2}$. This means the compressive stress is 33·5% of the small-deflection bending stress at the plate centre in the same direction.

From the foregoing definitions, the large-deflection relationships for the tabular data given may be summarised as follows:

$$W = K_l K_r Q \qquad (A6.2)$$

$$\bar{\sigma}_b = K_l K_r Q \qquad (A6.3)$$

$$\bar{\sigma}_m = K_l K_r Q \qquad (A6.4)$$

Thus, for any given value of transverse pressure Q, the magnitudes of the coefficients K_l and K_r are read directly from the appropriate data table and the required value of deflection or stress is then given by the multiple of these coefficients and Q. As an example, consider the square simply supported plate discussed in the preceding examples.

For $Q = 300$, Table A1 gives

$$W = K_l K_r Q = 4.434 \times 10^{-2} \times 0.304 \times 300 = 4.04$$

$$w = 4.04t$$

$$\bar{\sigma}_{bx1} = \bar{\sigma}_{by1} = K_l K_r Q = 0.168 \times 0.286 \times 300 = 14.41$$

hence

$$\sigma_{bx1} = \sigma_{by1} = (t/a)^2 E \bar{\sigma}_{bx1} = 14.41(t/a)^2 E$$

For membrane stresses at centre

$$\bar{\sigma}_{mx1} = \bar{\sigma}_{my1} = K_l K_r Q = 0.117 \times 0.286 \times 300 = 10.04$$

hence

$$\sigma_{mx1} = \sigma_{my1} = 10.04(t/a)^2 E$$

Central total stresses at the no-load surface are

$$\bar{\sigma}_{x1} = \bar{\sigma}_{bx1} + \bar{\sigma}_{mx1} = 14.41 + 10.04 = 24.45$$

$$\bar{\sigma}_{y1} = \bar{\sigma}_{x1} = 24.45$$

Alternatively, the total stresses may be obtained directly by adding the related coefficients, thus

$$\sigma_{x1} = (0.168 + 0.117) \times 300 \times 0.286(t/a)^2 E = 24.45(t/a)^2 E$$

For the point $x = a/2$, $y = 0$ (denoted as point 2 in the data tables)

$$\sigma_{bx2} = \sigma_{by2} = 0$$

$$\bar{\sigma}_{mx2} = 0, \text{ by definition}$$

$$\bar{\sigma}_{my2} = K_l K_r Q = -0.335 \times 0.286 \times 300 = -28.7$$

Due to symmetry, the flexural and membrane shearing stresses are both zero at the points discussed.

A7 Shearing stresses τ_b, τ_m

The points for which the state of stress is defined in the data tables are located either on the centre lines or on the boundaries, so that due to symmetry or the postulated boundary conditions, membrane shearing stresses τ_m (due to N_{xy}) as well as bending shearing stresses τ_b (due to M_{xy}) are zero. Hence the algebraic sum of the appropriate bending and membrane stresses at the points discussed gives the values of maximum and minimum surface stresses at these points. Furthermore, the stress components given suffice for the determination of equivalent stresses at the points considered.

A8 Signs

The signs of both the small-deflection and large-deflection components of deflections and stresses are obtained from the products of the appropriate coefficients given in the data tables.

For deflections, positive sign indicates transverse movement in direction of applied loading.

For bending stresses, positive sign indicates tensile and negative sign compressive stresses at the no-load surface (i.e., signs reverse on the loaded surface).

For membrane stresses, positive sign stands for tension and negative sign for compression.

A9 Poisson's ratio

The coefficients given are all evaluated for a Poisson's ratio of 0·3. The corresponding coefficients for values of Poisson's ratio other than 0·3 may be calculated using the following relationships, which are exact for cases of stress-free membrane boundary conditions (such as the plate problems of Table A1), and are approximate for other cases. In the following suffixes $v1$ and $v2$ relate to Poisson's ratios $v1$ and $v2$.

For deflections

$$W_{v2} = [(1 - v2^2)/(1 - v1^2)]W_1 \qquad (A9.1)$$

hence, for $v1 = 0·3$

$$W_{v2} = 1·099(1 - v2^2)W_1$$

If the given bending stresses are for a Poisson's ratio of $v1$ and it is required to evaluate the relating bending stresses for a material with a Poisson's ratio of $v2$, the following relationships apply:

$$\sigma_{bx,v2} = \frac{1}{1-v1^2}[(1 - v1v2)\sigma_{bx,v1} + (v2 - v1)\sigma_{by,v1}] \quad (A9.2)$$

$$\sigma_{by,v2} = \frac{1}{1-v1^2}[(1 - v1v2)\sigma_{by,v1} + (v2 - v1)\sigma_{bx,v1}] \quad (A9.3)$$

For $v1 = 0.3$, which applies to this data tabulated in this volume,

$$\sigma_{bx}^* = 1.1\,|(1 - 0.3v^*)\sigma_{bx} + (v^* - 0.3)\sigma_{by}| \quad (A9.4)$$

$$\sigma_{by}^* = 1.1\,|(1 - 0.3^*)\sigma_{by} + (v^* - 0.3)\sigma_{bx}| \quad (A9.5)$$

For membrane stresses

$$\sigma_{mx}^* \simeq \sigma_{mx} \quad (A9.6)$$

$$\sigma_{my}^* \simeq \sigma_{my} \quad (A9.7)$$

A10 Summary of some useful relationships for use with data tables

(1) Small deflection—linear behaviour

For transverse deflection

$$w = K_l Qt = K_l(qa^4/Et^3) \quad (A5.2)$$

For bending stresses

$$\sigma_b = K_l(t/a)^2 EQ = K_l(a/t)^2 q \quad (A5.4)$$

For membrane stresses

$$\sigma_m = 0$$

(2) Large-deflection behaviour

$$W = K_l K_r Q$$
$$w = K_l K_r Qt \quad (A6.2)$$

$$\bar{\sigma}_b = K_l K_r Q$$
$$\sigma_b = K_l K_r (t/a)^2 EQ \qquad (A6.3)$$

$$\bar{\sigma}_m = K_l K_r Q$$
$$\sigma_m = K_l K_r (t/a)^2 EQ \qquad (A6.4)$$

(3) Total stresses
At the loaded surface
$$\sigma_x = -\sigma_{bx} + \sigma_{mx}$$
$$\sigma_y = -\sigma_{by} + \sigma_{my} \qquad (A10.1)$$

At no-load surface
$$\sigma_x = \sigma_{bx} + \sigma_{mx}$$
$$\sigma_y = \sigma_{by} + \sigma_{my} \qquad (A10.2)$$

(4) Equivalent stresses
$$\sigma_e = (\sigma_x^2 + \sigma_y^2 - \sigma_x \sigma_y)^{\frac{1}{2}} \qquad (A10.3)$$

(5) Poisson's ratio
For a Poisson's ratio v other than 0·3
$$W_v = 1 \cdot 1(1 - v^2)W \qquad (A9.1)$$
$$\sigma_{bx,v} = 1 \cdot 1(1 - 0 \cdot 3v)\sigma_{bx} + (v - 0 \cdot 3)\sigma_{by} \qquad (A9.4)$$
$$\sigma_{by,v} = 1 \cdot 1(1 - 0 \cdot 3v)\sigma_{by} + (v - 0 \cdot 3)\sigma_{bx} \qquad (A9.5)$$
$$\sigma_{mx,v} \simeq \sigma_{mx} \qquad (A9.6)$$
$$\sigma_{my,v} \simeq \sigma_{my} \qquad (A9.7)$$

When dealing with concentrated loadings, in all the aforementioned expressions of this section, Q is to be replaced by the value of total loading P.

142 Thin Plate Design For Transverse Loading

A11 Additional numerical examples

Example A11.1   Steel plate under uniform transverse loading

Evaluate the central deflection and stresses of the following steel plate under water pressure.

Dimensions:

$a = 1000$ mm (39·4 in)
$b = 1000$ mm (39·4 in)
$t = 8$ mm (0·31 in)

Material:

High tensile steel, $E = 2·06 \times 10^5$ N/mm² (30 000 ksi), Poisson's ratio 0·3.

Boundary conditions:

FBC: rigidly supported, rotationally free on all four sides.
MBC: membrane direct and shear stresses zero on all sides.

Loading:

Uniformly distributed water pressure of intensity $q = 0·253$ N/mm² (36·7 psi)

Boundary conditions and loading relate to data Table A1 as shown in Fig. A2(i).

$b/a = 1000/1000 = 1$

$Q = a^4 q / t^4 E = 1000^4 \times 0·253 / 8^4 \times 2·06 \times 10^5 = 300$

The coefficients for this loading are given in the first section of Table A1 on the $Q = 300$ row.

(a) Small deflection (linear)
    Central deflection

$w = K_l Q t = 4·434 \times 10^{-2} \times 300 \times 8 = 106$ mm (4·19 in)

Central stresses

$\sigma_{bx1} = \sigma_{by1} = K_l (a/t)^2 q = 0·286 \times (1000/8)^2 \times 0·253$
$= 1130$ N/mm² (164 ksi)

Obviously these values will not be reached in practice due to the plate's large-deflection behaviour. $\sigma_{mx1} = \sigma_{my1} = 0$

(b) Large deflection
Central deflection

$$w = K_l K_r Q t = 4 \cdot 434 \times 10^{-2} \times 0 \cdot 304 \times 300 \times 8 = 32 \cdot 3 \text{ mm}$$
$$(1 \cdot 27 \text{ in})$$

Central stresses

$$\sigma_{bx1} = \sigma_{by1} = K_l K_r (t/a)^2 EQ$$
$$= 0 \cdot 286 \times 0 \cdot 168 \times (8/1000)^2 \times 2 \cdot 06 \times 10^5 \times 300$$
$$= 190 \text{ N/mm}^2 \text{ (27·6 ksi)}$$

$$\sigma_{mx1} = \sigma_{my1} = K_l K_r (t/a)^2 EQ$$
$$= 0 \cdot 286 \times 0 \cdot 117 \times (8/1000)^2 \times 2 \cdot 06 \times 10^5 \times 300$$
$$= 132 \text{ N/mm}^2 \text{ (19·2 ksi)}$$

or simply

$$\sigma_{mx1} = \sigma_{my1} = 0 \cdot 117 \times 190/0 \cdot 168 = 132 \text{ N/mm}^2 \text{ (19·2 ksi)}$$

Central stresses at no-load surface

$$\sigma_x = \sigma_y = 190 + 132 = 322 \text{ N/mm}^2 \text{ (46·8 ksi)}$$

Central stress at loaded surface

$$\sigma_x = \sigma_y = -190 + 132 = -58 \text{ N/mm}^2 \text{ (−8·4 ksi) compressive}$$

Example A11.2 Outer shell of an aircraft wing

Consider the case of a man stepping on the middle of a panel on the outer shell of an aircraft wing.

Dimensions:

$a = 200$ mm (7·87 in)
$b = 400$ mm (15·7 in)
$t = 1 \cdot 6$ mm (0·063 in)

Material:

Aluminium alloy with $E = 8 \times 10^4$ N/mm² (11 600 ksi), proof stress 480 N/mm² (70 ksi), Poisson's ratio 0·316 (assumed equal to 0·3 herein).

144 Thin Plate Design For Transverse Loading

Boundary conditions:

The panel is continuous over stiffeners and webs in both directions. The adjacent panels may be assumed to be of identical dimensions and hence the flexural and membrane boundary conditions correspond to the case of a continuous plate.

Loading:

Total weight of man $P = 750$ N (169 lb). While stepping, the load is assumed to be distributed uniformly over a contact area of dimensions 40 × 65 mm (1·57 × 2·56 in).

From Fig. A2(ii), the boundary conditions and loading correspond to the data given in Table A21.

For maximum deflection, the loading is assumed to be central with its longer sides parallel to the longitudinal axis of the panel.

$b/a = 400/200 = 2$

$\alpha = u/a = 40/200 = 0·2$

$\beta = v/a = 65/200 = 0·325$

The dimensions of the patch loading $\alpha \times \beta$ may be safely approximated as 0·2 × 0·3, for which the coefficients in the fourth section of Table A21 apply.

$P = a^2 p/t^4 E = 200^2 \times 750/1·6^4 \times 8 \times 10^4 = 57·2$

The required coefficients are to be interpolated between values given for loadings 20 and 60.

Central deflection:

$w = K_l K_r P t$

$K_r = 0·344 + \dfrac{60 - 57·2}{60 - 20}(0·597 - 0·344) = 0·362$

$w = 0·1110 \times 0·362 \times 57·2 \times 1·6 = 3·68$ mm (0·145 in)

Central stresses

For σ_{bx1}

$K_r = 0·394 + \dfrac{60 - 57·2}{60 - 20}(0·638 - 0·394) = 0·394 + 0·07 \times 0·244$

$= 0·411$

$$\sigma_{bx1} = K_l K_r (t/a)^2 EP = 1\cdot18 \times 0\cdot411 \times (1\cdot6/200)^2 \times 8 \times 10^4$$
$$\times 57\cdot2 = 142 \text{ N/mm}^2 \text{ (20·6 ksi)}$$

For σ_{by1}, $K_r = 0\cdot375 + 0\cdot07(0\cdot627 - 0\cdot375) = 0\cdot393$

$$\sigma_{by1} = 0\cdot897 \times 0\cdot392 \times (1\cdot6/200)^2 \times 8 \times 10^4 \times 57\cdot2$$
$$= 103 \text{ N/mm}^2 \text{ (14·9 ksi)}$$

$\sigma_{mx1} = 142 \times 0\cdot136/0\cdot394 = 49 \text{ N/mm}^2 \text{ (7·11 ksi)}$

Similarly

$\sigma_{my1} = 103 \times 0\cdot137/0\cdot375 = 37 \text{ N/mm}^2 \text{ (5·46 ksi)}$

Maximum stress occurs on no-load surface in the transverse direction

$\sigma_{max} = \sigma_x = \sigma_{bx} + \sigma_{mx} = 142 + 49 = 191 \text{ N/mm}^2 \text{ (27·71 ksi)}$

For equivalent stress

$\sigma_y = 103 + 37 = 140 \text{ N/mm}^2 \text{ (20·36 ksi)}$
$\sigma_e = (\sigma_x^2 + \sigma_y^2 - \sigma_x \sigma_y)^{\frac{1}{2}} = (191^2 + 140^2 - 191 \times 140)^{\frac{1}{2}}$
$= 171 \text{ N/mm}^2 \text{ (24·84 ksi)}$

Example A11.3 Stresses on the boundaries

Calculate the central deflection and the stresses at mid-point of the longer edges of the following steel plate.

Dimensions:

$a = 300$ mm (11·8 in)
$b = 500$ mm (19·7 in)
$t = 4$ mm (0·157 in)

Material:

High tensile steel with $E = 2\cdot06 \times 10^5$ N/mm^2 (30 000 ksi), Poisson's ratio 0·3.

146 Thin Plate Design For Transverse Loading

Boundary conditions:

FBC: Longer edges rigidly supported, rotationally fixed; shorter edges rigidly supported, rotationally free.
MBC: Longer edges held in position with membrane shear stresses equal to zero; for shorter edges zero membrane direct and shear stresses.

Loading:

UDL with intensity $q = 0.13$ N/mm² (18.1 psi)

The specified boundary conditions and loading refer to Fig. A2(iii) and data Table A28.

$b/a = 500/300 = 1.67$

The aspect ratio 1·67 may be approximated, for design purposes, to be 1·5, for which data is given in the first section of Table A28. If a more exact analysis is required, the calculated deflections and stresses may be corrected for the aspect ratio, and a better approximation made, as is illustrated in this example.

$Q = a^4 q / t^4 E = 300^4 \times 0.13/4^4 \times 2.06 \times 10^5 = 20$

Central deflection

$w = K_l K_r Q t = 0.027\,06 \times 0.901 \times 29 \times 4 = 1.95$ mm

(0.0768 in)

Stresses at mid-point of the longer edge—point 2 ($x = a/2$, $y = 0$)

$\sigma_{bx2} = K_l K_r (t/a)^2 EQ = -0.495 \times 0.936 \times (4/300)^2 \times 2.06 \times 10^5$
$\times 20 = -339$ N/mm² (-49.2 ksi)

$\sigma_{by2} = \nu \sigma_{bx2} = -0.3 \times 339 = -102$ N/mm² (-14.8 ksi)

$\sigma_{mx2} = 339 \times 0.053/0.936 = 19.2$ N/mm² (2.78 ksi)

$\sigma_{my2} = -0.148 \times 0.046 \times (4/300)^2 \times 2.06 \times 10^5 \times 20$
$= -4.99$ N/mm² (0.72 ksi)

Maximum direct stress occurs on the loaded surface in the x-direction

$\sigma_{max} = 339 + 19.2 = 358.2$ N/mm² (51.7 ksi)

Maximum equivalent stress occurs on the loaded surface, for which

$\sigma_x = 358 \cdot 2 \text{ N/mm}^2 \text{ (51·7 ksi)}$

$\sigma_y = 102 - 4 \cdot 99 = 97 \cdot 01 \text{ N/mm}^2 \text{ (14·1 ksi)}$

$\sigma_e = (358 \cdot 2^2 + 97 \cdot 01^2 - 358 \cdot 2 \times 97 \cdot 01)^{\frac{1}{2}} = 321 \text{ N/mm}^2$

(46·5 ksi)

Adjustment for correct aspect ratio:

First the corresponding deflections and stresses are evaluated for a plate with the same boundary conditions and loading, but with $b/a = 2$. Using a similar procedure and the second section of data Table A28:

For $b/a = 2$

$w \quad = 2 \cdot 00 \text{ mm } (0 \cdot 0789 \text{ in})$

$\sigma_{max} = 363 \text{ N/mm}^2 \text{ (52·6 ksi)}$

$\sigma_e \quad = 325 \text{ N/mm}^2 \text{ (47·1 ksi)}$

The adjustment may then be carried out for aspect ratio 1·67 using linear interpolation:

$w = 1 \cdot 95 + \dfrac{1 \cdot 67 - 1 \cdot 5}{2 - 1 \cdot 5}(2 \cdot 00 - 1 \cdot 95) = 1 \cdot 967 \text{ mm } (0 \cdot 0774 \text{ in})$

$\sigma_{max} = 358 \cdot 2 + \dfrac{1 \cdot 67 - 1 \cdot 5}{2 - 1 \cdot 5}(363 - 358 \cdot 2) = 358 \cdot 9 \text{ N/mm}^2$

(52·18 ksi)

Similarly, the adjusted value of σ_e equals 322·4 N/mm² (46·7 ksi). In fact, the variations of deflections and stresses with aspect ratio are not linear, but the procedure adopted in the preceding example is accurate enough for design purposes.

Example A11.4 Variations in Poisson's ratio

Consider an open cubic container with glass walls holding a liquid with specific gravity of 0·878. Calculate the deflections and stresses of the unsupported edges at the top.

148 Thin Plate Design For Transverse Loading

Dimensions:

$a = b = 600$ mm (23·6 in)

$t = 3$ mm (0·118 in)

Material:

Glass with $E = 6·89 \times 10^4$ N/mm² (10 000 ksi), Poisson's ratio 0·22.

Boundary conditions:

The fixtures are such that the sides and the bottom may be assumed as rigidly supported, rotationally free, and the top edge as rotationally free with no support; for the membrane boundary conditions, the sides and bottom edges remain straight with the integral of the in-plane direct stresses along each equal to zero together with membrane shear stresses zero. There are no membrane stresses at the top edge.

Loading:

Linearly varying distribution of pressure, zero at top and maximum intensity at bottom equal to

$q = 0·878 \times 9·806 \times 10^{-6} \times 600 = 5·166 \times 10^{-3}$ N/mm²

(0·749 psi)

(weight of water $9·806 \times 10^{-6}$ N/mm³)

From Fig. A2(iv), the boundary conditions and loading given refer to the second section of Table A37. The required values are first evaluated from the data tables for a Poisson's ratio of 0·3, and subsequently adjusted for the correct material Poisson's ratio.

$b/a = 600/600 = 1$

$Q = a^4 q_{max}/t^4 E = 600^4 \times 5·166 \times 10^{-3}/3^4 \times 6·89 \times 10^4 = 120$

Deflection at mid-point of the free edge

$w_3 = K_t K_r Q t = 2·315 \times 10^{-2} \times 0·5 \times 120 \times 3 = 4·17$ mm

(0·164 in)

Stresses at mid-point of the free edge

$\sigma_{bx3} = K_l K_r (t/a)^2 EQ = 1 \cdot 13 \times 10^{-1} \times 0 \cdot 505 \times (3/600)^2$
$\qquad \times 6 \cdot 89 \times 10^4 \times 120 = 11 \cdot 8 \text{ N/mm}^2 \text{ (1710 psi)}$

$\sigma_{mx3} = -11 \cdot 8 \times 0 \cdot 018/0 \cdot 505 = -0 \cdot 420 \text{ N/mm}^2 \text{ (61 psi)}$

$\sigma_{max} = 11 \cdot 8 - 0 \cdot 420 = 11 \cdot 38 \text{ N/mm}^2$ (1649 psi) (maximum tensile stress)

$\sigma_{by3} = 0 = \sigma_{my3}$

Adjustment for Poisson's ratio of 0·22

$w_3 = 1 \cdot 1 \times (1 - 0 \cdot 22^2) \times 4 \cdot 17 = 4 \cdot 36 \text{ mm} \text{ (0·172 in)}$

$\sigma_{bx3} = 1 \cdot 1 \times (1 - 0 \cdot 3 \times 0 \cdot 22) \times 11 \cdot 8 = 12 \cdot 1 \text{ N/mm}^2 \text{ (1758 psi)}$

$\sigma_{mx3} = -0 \cdot 420 \text{ N/mm}^2 \text{ (61 psi)}$

maximum tensile stress

$\sigma_{max} = 12 \cdot 1 - 0 \cdot 420 = 11 \cdot 68 \text{ N/mm}^2 \text{ (1694 psi)}$

Table A1. Large-deflection Reduction Coefficients K_r

LOADING : UDL   intensity q

$W = K_\ell K_r Q$, $\sigma = (t/a)^2 \cdot E\bar{\sigma}$ $W = w/t$

$\bar{\sigma} = K_\ell K_r Q$, $Q = a^4 q / t^4 E$

b/a	Q	W_1	$\bar{\sigma}_{bx1}$	$\bar{\sigma}_{by1}$	$\bar{\sigma}_{mx1}$	$\bar{\sigma}_{my1}$	$\bar{\sigma}_{my2}$
1	Linear K_ℓ	4.434 x10^{-2}	0.286	0.286	0.286	0.286	0.286
	20	0.894	0.850	0.850	0.106	0.106	-0.206
	40	0.753	0.662	0.662	0.141	0.141	-0.294
	120	0.482	0.336	0.336	0.144	0.144	-0.358
	200	0.375	0.230	0.230	0.130	0.130	-0.351
	300	0.304	0.168	0.168	0.117	0.117	-0.335
	400	0.261	0.134	0.134	0.107	0.107	-0.320
1.5	Linear K_ℓ	8.438 x10^{-2}	0.486	0.299	0.486	0.299	0.299
	20	0.812	0.764	0.721	0.050	0.232	-0.416
	40	0.647	0.570	0.517	0.049	0.275	-0.499
	120	0.394	0.298	0.270	0.035	0.264	-0.491
	200	0.304	0.206	0.195	0.029	0.239	-0.456
	300	0.245	0.149	0.147	0.026	0.216	-0.425
	400	0.209	0.117	0.120	0.024	0.199	-0.403
2	Linear K_ℓ	11.07 x10^{-2}	0.609	0.278	0.609	0.278	0.278
	20	0.832	0.806	0.779	0.014	0.288	-0.484
	40	0.674	0.630	0.607	0.013	0.354	-0.596
	120	0.417	0.351	0.345	0.012	0.357	-0.614
	200	0.322	0.249	0.250	0.011	0.329	-0.579
	300	0.259	0.184	0.190	0.010	0.302	-0.543
	400	0.221	0.145	0.154	0.010	0.281	-0.516
3	Linear K_ℓ	13.42 x10^{-2}	0.712	0.244	0.712	0.244	0.244
	20	0.932	0.931	0.970	-0.001	0.261	-0.417
	40	0.806	0.793	0.832	0.001	0.389	-0.629
	60	0.709	0.687	0.726	0.004	0.447	-0.730

Appendix 151

Table A2. Large-deflection Reduction Coefficients K_r

LOADING : UDL intensity q

$W = K_\ell K_r Q$, $\quad \sigma = (t/a)^2 \cdot E\bar{\sigma}$, $\quad W = w/t$

$\bar{\sigma} = K_\ell K_r Q$, $\quad Q = a^4 q/t^4 E$

b/a	Q	W_1	$\bar{\sigma}_{bx1}$	$\bar{\sigma}_{by1}$	$\bar{\sigma}_{mx1}$	$\bar{\sigma}_{my1}$	$\bar{\sigma}_{mx2}$	$\bar{\sigma}_{my2}$
1	Linear K_ℓ ➡	4.434 ×10⁻²	0.286	0.286	0.286	0.286	0.286	0.286
	20	0.666	0.635	0.635	0.184	0.184	0.193	0.038
	40	0.481	0.440	0.440	0.193	0.193	0.207	0.043
	120	0.255	0.215	0.215	0.166	0.166	0.183	0.043
	200	0.185	0.150	0.150	0.147	0.147	0.165	0.041
	300	0.143	0.112	0.112	0.132	0.132	0.149	0.039
	400	0.118	0.091	0.091	0.122	0.122	0.138	0.037
1.5	Linear K_ℓ ➡	8.438 ×10⁻²	0.486	0.299	0.486	0.299	0.486	0.299
	20	0.476	0.449	0.404	0.185	0.176	0.194	0.073
	40	0.322	0.293	0.249	0.171	0.165	0.181	0.075
	120	0.162	0.140	0.108	0.131	0.131	0.141	0.067
	200	0.116	0.098	0.074	0.114	0.114	0.122	0.062
	300	0.089	0.074	0.055	0.100	0.102	0.108	0.057
	400	0.073	0.061	0.044	0.092	0.094	0.099	0.053
2	Linear K_ℓ ➡	11.07 ×10⁻²	0.609	0.278	0.609	0.278	0.609	0.278
	20	0.388	0.364	0.298	0.166	0.156	0.172	0.104
	40	0.258	0.235	0.179	0.148	0.144	0.154	0.103
	120	0.127	0.110	0.079	0.110	0.112	0.115	0.087
	200	0.091	0.078	0.055	0.095	0.098	0.099	0.078
	300	0.069	0.059	0.041	0.084	0.088	0.088	0.071
	400	0.057	0.048	0.034	0.076	0.081	0.080	0.065
3	Linear K_ℓ ➡	13.42 ×10⁻²	0.712	0.244	0.712	0.244	0.712	0.244
	20	0.323	0.306	0.267	0.144	0.145	0.146	0.138
	40	0.212	0.196	0.170	0.126	0.132	0.127	0.128
	120	0.105	0.093	0.081	0.095	0.105	0.095	0.103
	200	0.075	0.066	0.057	0.082	0.093	0.082	0.091
	300	0.057	0.050	0.043	0.072	0.083	0.073	0.081
	400	0.047	0.041	0.036	0.066	0.076	0.066	0.075

Table A3. Large-deflection Reduction Coefficients K_r

LOADING : UDL intensity q

$W = K_\ell K_r Q,$ $\sigma = (t/a)^2 . E\bar{\sigma}$ $W = w/t$

$\bar{\sigma} = K_\ell K_r Q,$ $Q = a^4 q / t^4 E$

b/a	Q	W_1	$\bar{\sigma}_{bx1}$	$\bar{\sigma}_{by1}$	$\bar{\sigma}_{mx1}$	$\bar{\sigma}_{my1}$
1	Linear K_ℓ	4.434 x10^{-2}	0.286	0.286	0.286	0.286
	20	0.812	0.765	0.765	0.138	0.138
	40	0.637	0.558	0.558	0.165	0.165
	120	0.372	0.271	0.271	0.155	0.155
	200	0.279	0.185	0.185	0.139	0.139
	300	0.220	0.135	0.135	0.125	0.125
	400	0.186	0.108	0.108	0.115	0.115
1.5	Linear K_ℓ	8.438 x10^{-2}	0.486	0.299	0.486	0.299
	20	0.683	0.636	0.584	0.072	0.276
	40	0.513	0.453	0.402	0.072	0.303
	120	0.287	0.224	0.197	0.055	0.269
	200	0.214	0.154	0.139	0.047	0.240
	300	0.168	0.113	0.105	0.041	0.217
	400	0.142	0.090	0.086	0.037	0.200
2	Linear K_ℓ	11.07 x10^{-2}	0.609	0.278	0.609	0.278
	20	0.698	0.670	0.628	0.033	0.386
	40	0.522	0.485	0.452	0.031	0.419
	120	0.296	0.252	0.240	0.025	0.378
	200	0.221	0.176	0.172	0.022	0.341
	300	0.174	0.130	0.130	0.020	0.309
	400	0.146	0.104	0.106	0.018	0.286
3	Linear K_ℓ	13.42 x10^{-2}	0.712	0.244	0.712	0.244
	20	0.810	0.804	0.831	0.006	0.452
	40	0.641	0.625	0.651	0.009	0.556
	120	0.379	0.351	0.368	0.010	0.558
	200	0.286	0.254	0.268	0.010	0.518
	300	0.226	0.192	0.205	0.009	0.477
	400	0.191	0.156	0.167	0.008	0.447

Appendix   153

Table A4. Large-deflection Reduction Coefficients K_r

LOADING : UDL   intensity q

$W = K_\ell K_r Q$, $\bar{\sigma} = (t/a)^2 . E\bar{\sigma}$ $W = w/t$

$\bar{\sigma} = K_\ell K_r Q$, $Q = a^4 q/t^4 E$

b/a	Q	W_1	$\bar{\sigma}_{bx1}$	$\bar{\sigma}_{by1}$	$\bar{\sigma}_{mx1}$	$\bar{\sigma}_{my1}$	$\bar{\sigma}_{mx2}$	$\bar{\sigma}_{my2}$
1	Linear K_ℓ ➡	4.434 ×10⁻²	0.286	0.286	0.286	0.286	0.286	0.286
	20	0.833	0.801	0.801	0.110	0.110	0.123	-0.120
	40	0.661	0.604	0.604	0.134	0.134	0.159	-0.150
	120	0.383	0.310	0.310	0.126	0.126	0.167	-0.148
	200	0.285	0.217	0.217	0.111	0.111	0.155	-0.133
	300	0.223	0.163	0.163	0.098	0.098	0.141	-0.118
	400	0.186	0.134	0.134	0.089	0.089	0.130	-0.108
1·5	Linear K_ℓ ➡	8.438 ×10⁻²	0.486	0.299	0.486	0.299	0.486	0.299
	20	0.656	0.622	0.539	0.127	0.096	0.145	-0.099
	40	0.474	0.435	0.335	0.120	0.095	0.146	-0.098
	120	0.254	0.222	0.143	0.086	0.075	0.115	-0.075
	200	0.186	0.160	0.099	0.071	0.064	0.097	-0.063
	300	0.145	0.124	0.075	0.061	0.056	0.084	-0.054
	400	0.122	0.104	0.063	0.055	0.051	0.076	-0.049
2	Linear K_ℓ ➡	11.07 ×10⁻²	0.609	0.278	0.609	0.278	0.609	0.278
	20	0.561	0.531	0.399	0.103	0.051	0.118	-0.051
	40	0.396	0.366	0.248	0.087	0.043	0.105	-0.043
	120	0.211	0.191	0.121	0.060	0.031	0.075	-0.030
	200	0.155	0.139	0.089	0.050	0.027	0.063	-0.024
	300	0.122	0.108	0.070	0.044	0.024	0.055	-0.021
	400	0.102	0.090	0.059	0.040	0.023	0.050	-0.018
3	Linear K_ℓ ➡	13.42 ×10⁻²	0.712	0.244	0.712	0.244	0.712	0.244
	20	0.511	0.493	0.419	0.067	0.008	0.071	-0.003
	40	0.365	0.347	0.297	0.059	0.008	0.062	-0.000
	120	0.199	0.185	0.160	0.045	0.011	0.047	0.004
	200	0.146	0.134	0.116	0.038	0.018	0.040	0.011
	300	0.114	0.103	0.089	0.034	0.024	0.036	0.018
	400	0.095	0.085	0.073	0.031	0.027	0.033	0.021

Table A5. Large-deflection Reduction Coefficients K_r

LOADING : UDL   intensity q

$W = K_\ell K_r Q,$ $\sigma = (t/a)^2 . E\bar{\sigma}$ $W = w/t$

$\bar{\sigma} = K_\ell K_r Q,$ $Q = a^4 q / t^4 E$

b/a	Q	W_1	$\bar{\sigma}_{bx1}$	$\bar{\sigma}_{by1}$	$\bar{\sigma}_{mx1}$	$\bar{\sigma}_{my1}$	$\bar{\sigma}_{bx2}$	$\bar{\sigma}_{my2}$
1	Linear K_ℓ ➡	1.375×10^{-2}	1.36×10^{-1}	1.36×10^{-1}	1.36×10^{-1}	1.36×10^{-1}	-3.08×10^{-1}	0.924×10^{-1}
	20	0.988	0.982	0.982	0.045	0.045	0.994	-0.079
	40	0.956	0.932	0.932	0.083	0.083	0.976	-0.148
	120	0.789	0.686	0.686	0.156	0.156	0.883	-0.310
	200	0.666	0.523	0.523	0.171	0.171	0.807	-0.376
	300	0.567	0.402	0.402	0.171	0.171	0.738	-0.414
	400	0.499	0.328	0.328	0.166	0.166	0.686	-0.432
1.5	Linear K_ℓ ➡	2.393×10^{-2}	2.18×10^{-1}	1.21×10^{-1}	2.18×10^{-1}	1.21×10^{-1}	-4.54×10^{-1}	1.36×10^{-1}
	20	0.979	0.971	0.960	0.028	0.110	0.985	-0.133
	40	0.927	0.901	0.868	0.045	0.193	0.947	-0.237
	120	0.722	0.645	0.575	0.057	0.318	0.791	-0.411
	200	0.602	0.506	0.444	0.053	0.343	0.696	-0.456
	300	0.509	0.403	0.355	0.048	0.344	0.620	-0.468
	400	0.448	0.338	0.301	0.045	0.338	0.569	-0.467
2	Linear K_ℓ ➡	2.763×10^{-2}	2.45×10^{-1}	0.945×10^{-1}	2.45×10^{-1}	0.945×10^{-1}	-4.98×10^{-1}	1.49×10^{-1}
	20	0.989	0.987	0.986	0.004	0.114	0.991	-0.101
	40	0.959	0.951	0.949	0.007	0.210	0.967	-0.186
	120	0.807	0.776	0.783	0.013	0.415	0.848	-0.370
	200	0.694	0.644	0.657	0.016	0.483	0.759	-0.434
	300	0.598	0.533	0.548	0.017	0.511	0.682	-0.463
	400	0.531	0.457	0.472	0.018	0.515	0.629	-0.471
3	Linear K_ℓ ➡	2.870×10^{-2}	2.48×10^{-1}	0.754×10^{-1}	2.48×10^{-1}	0.754×10^{-1}	-5.05×10^{-1}	1.52×10^{-1}
	20	1.003	1.005	1.017	-0.006	0.036	1.003	-0.022
	40	1.009	1.015	1.054	-0.009	0.087	1.009	-0.055
	120	0.979	0.984	1.077	-0.004	0.328	0.987	-0.218
	200	0.906	0.899	0.995	-0.000	0.488	0.931	-0.327
	250	0.861	0.845	0.938	0.001	0.551	0.895	-0.371

Appendix 155

Table A6. Large-deflection Reduction Coefficients K_r

LOADING : UDL   intensity q

$W = K_\ell K_r Q$, $\sigma = (t/a)^2 . E\bar{\sigma}$ $W = w/t$

$\bar{\sigma} = K_\ell K_r Q$, $Q = a^4 q / t^4 E$

b/a	Q	W_1	$\bar{\sigma}_{bx1}$	$\bar{\sigma}_{by1}$	$\bar{\sigma}_{mx1}$	$\bar{\sigma}_{my1}$	$\bar{\sigma}_{bx2}$	$\bar{\sigma}_{mx2}$	$\bar{\sigma}_{my2}$
1	Linear K_ℓ ▶	1.375×10^{-2}	1.36×10^{-1}	1.36×10^{-1}	1.36×10^{-1}	1.36×10^{-1}	-3.08×10^{-1}	3.08×10^{-1}	0.924×10^{-1}
	20	0.967	0.956	0.956	0.083	0.083	0.980	0.026	0.004
	40	0.892	0.860	0.860	0.140	0.140	0.935	0.045	0.006
	120	0.639	0.550	0.550	0.210	0.210	0.773	0.073	0.002
	200	0.504	0.400	0.400	0.215	0.215	0.677	0.079	0.003
	300	0.406	0.301	0.301	0.208	0.208	0.602	0.080	0.008
	400	0.346	0.243	0.243	0.200	0.200	0.551	0.079	0.012
1.5	Linear K_ℓ ▶	2.393×10^{-2}	2.18×10^{-1}	1.21×10^{-1}	2.18×10^{-1}	1.21×10^{-1}	-4.54×10^{-1}	4.54×10^{-1}	1.36×10^{-1}
	20	0.908	0.888	0.869	0.114	0.138	0.940	0.052	0.006
	40	0.770	0.723	0.682	0.163	0.196	0.848	0.075	0.014
	120	0.482	0.402	0.347	0.188	0.226	0.641	0.092	0.033
	200	0.365	0.284	0.235	0.179	0.215	0.547	0.090	0.040
	300	0.288	0.212	0.172	0.167	0.201	0.480	0.085	0.045
	400	0.242	0.172	0.137	0.158	0.190	0.436	0.081	0.047
2	Linear K_ℓ ▶	2.763×10^{-2}	2.45×10^{-1}	0.945×10^{-1}	2.45×10^{-1}	0.945×10^{-1}	-4.98×10^{-1}	4.98×10^{-1}	1.49×10^{-1}
	20	0.877	0.853	0.822	0.119	0.143	0.918	0.060	0.027
	40	0.718	0.670	0.615	0.159	0.191	0.809	0.082	0.044
	120	0.433	0.361	0.309	0.174	0.211	0.599	0.091	0.066
	200	0.325	0.254	0.213	0.165	0.200	0.510	0.087	0.070
	300	0.256	0.190	0.157	0.154	0.188	0.446	0.081	0.070
	400	0.215	0.154	0.127	0.146	0.178	0.405	0.076	0.070
3	Linear K_ℓ ▶	2.870×10^{-2}	2.48×10^{-1}	0.754×10^{-1}	2.48×10^{-1}	0.754×10^{-1}	-5.05×10^{-1}	5.05×10^{-1}	1.52×10^{-1}
	20	0.861	0.839	0.828	0.118	0.123	0.907	0.059	0.052
	40	0.701	0.656	0.644	0.156	0.167	0.796	0.079	0.074
	120	0.420	0.353	0.345	0.171	0.193	0.586	0.085	0.090
	200	0.316	0.249	0.244	0.162	0.188	0.497	0.081	0.089
	300	0.250	0.186	0.182	0.152	0.180	0.433	0.076	0.085
	400	0.210	0.151	0.148	0.144	0.172	0.392	0.072	0.082

Table A7. Large-deflection Reduction Coefficients K_r

LOADING : UDL   intensity q

$W = K_\ell K_r Q$, $\sigma = (t/a)^2 . E\bar{\sigma}$ $\bar{W} = w/t$

$\bar{\sigma} = K_\ell K_r Q$, $Q = a^4 q/t^4 E$

b/a	Q	$\bar{W}_1$	$\bar{\sigma}_{bx1}$	$\bar{\sigma}_{by1}$	$\bar{\sigma}_{mx1}$	$\bar{\sigma}_{my1}$	$\bar{\sigma}_{bx2}$
1	Linear K_ℓ ➡	1.375×10^{-2}	1.36×10^{-1}	1.36×10^{-1}	1.36×10^{-1}	1.36×10^{-1}	-3.08×10^{-1}
	20	0.980	0.972	0.972	0.059	0.059	0.988
	40	0.931	0.903	0.903	0.106	0.106	0.959
	120	0.722	0.622	0.622	0.179	0.179	0.827
	200	0.591	0.462	0.462	0.190	0.190	0.735
	300	0.490	0.351	0.351	0.187	0.187	0.659
	400	0.425	0.284	0.284	0.180	0.180	0.605
1.5	Linear K_ℓ ➡	2.393×10^{-2}	2.18×10^{-1}	1.21×10^{-1}	2.18×10^{-1}	1.21×10^{-1}	-4.54×10^{-1}
	20	0.961	0.950	0.935	0.044	0.148	0.971
	40	0.881	0.848	0.808	0.068	0.244	0.910
	120	0.633	0.556	0.489	0.085	0.357	0.718
	200	0.507	0.418	0.361	0.080	0.367	0.616
	300	0.417	0.324	0.280	0.074	0.359	0.541
	400	0.360	0.267	0.233	0.070	0.349	0.492
2	Linear K_ℓ ➡	2.763×10^{-2}	2.45×10^{-1}	0.945×10^{-1}	2.45×10^{-1}	0.945×10^{-1}	-4.98×10^{-1}
	20	0.974	0.969	0.961	0.016	0.180	0.979
	40	0.914	0.899	0.883	0.025	0.312	0.932
	120	0.698	0.655	0.644	0.036	0.518	0.760
	200	0.573	0.516	0.512	0.037	0.561	0.660
	300	0.478	0.411	0.413	0.037	0.567	0.581
	400	0.416	0.345	0.349	0.036	0.559	0.530
3	Linear K_ℓ ➡	2.870×10^{-2}	2.48×10^{-1}	0.754×10^{-1}	2.48×10^{-1}	0.754×10^{-1}	-5.05×10^{-1}
	20	1.001	1.003	1.021	-0.006	0.108	1.001
	40	0.997	1.002	1.052	-0.007	0.225	0.998
	120	0.890	0.882	0.965	0.005	0.571	0.915
	200	0.776	0.752	0.829	0.010	0.718	0.824
	250	0.718	0.687	0.759	0.012	0.764	0.778

Appendix 157

Table A8. Large-deflection Reduction Coefficients K_r

LOADING : UDL   intensity q

$W = K_\ell K_r Q$, $\bar{\sigma} = (t/a)^2 . E\bar{\sigma}$ $W = w/t$

$\bar{\sigma} = K_\ell K_r Q$, $Q = a^4 q / t^4 E$

b/a	Q	W_1	$\bar{\sigma}_{bx1}$	$\bar{\sigma}_{by1}$	$\bar{\sigma}_{mx1}$	$\bar{\sigma}_{my1}$	$\bar{\sigma}_{bx2}$	$\bar{\sigma}_{mx2}$	$\bar{\sigma}_{my2}$
1	Linear K_ℓ ➡	1.375×10^{-2}	1.36×10^{-1}	1.36×10^{-1}	1.36×10^{-1}	1.36×10^{-1}	-3.08×10^{-1}	3.08×10^{-1}	0.924×10^{-1}
	20	0.986	0.980	0.980	0.048	0.048	0.993	0.010	-0.060
	40	0.949	0.925	0.925	0.087	0.087	0.974	0.019	-0.111
	120	0.765	0.668	0.668	0.159	0.159	0.878	0.040	-0.219
	200	0.635	0.506	0.506	0.172	0.172	0.803	0.049	-0.252
	300	0.531	0.388	0.388	0.170	0.170	0.736	0.055	-0.264
	400	0.461	0.316	0.316	0.164	0.164	0.688	0.057	-0.266
1.5	Linear K_ℓ ➡	2.393×10^{-2}	2.18×10^{-1}	1.21×10^{-1}	2.18×10^{-1}	1.21×10^{-1}	-4.54×10^{-1}	4.54×10^{-1}	1.36×10^{-1}
	20	0.960	0.947	0.930	0.063	0.080	0.976	0.026	-0.060
	40	0.877	0.841	0.791	0.100	0.128	0.924	0.044	-0.098
	120	0.622	0.539	0.440	0.126	0.167	0.758	0.065	-0.136
	200	0.493	0.402	0.305	0.119	0.161	0.667	0.067	-0.134
	300	0.402	0.311	0.227	0.109	0.149	0.597	0.066	-0.127
	400	0.345	0.258	0.183	0.101	0.139	0.549	0.064	-0.120
2	Linear K_ℓ ➡	2.763×10^{-2}	2.45×10^{-1}	0.945×10^{-1}	2.45×10^{-1}	0.945×10^{-1}	-4.98×10^{-1}	4.98×10^{-1}	1.49×10^{-1}
	20	0.947	0.934	0.903	0.057	0.064	0.966	0.030	-0.035
	40	0.848	0.814	0.742	0.084	0.091	0.902	0.047	-0.052
	120	0.588	0.521	0.427	0.097	0.099	0.725	0.060	-0.058
	200	0.466	0.393	0.313	0.091	0.090	0.635	0.058	-0.053
	300	0.380	0.307	0.243	0.085	0.081	0.567	0.055	-0.048
	400	0.327	0.256	0.201	0.080	0.075	0.522	0.052	-0.044
3	Linear K_ℓ ➡	2.870×10^{-2}	2.48×10^{-1}	0.754×10^{-1}	2.48×10^{-1}	0.754×10^{-1}	-5.05×10^{-1}	5.05×10^{-1}	1.52×10^{-1}
	20	0.949	0.940	0.928	0.041	0.013	0.966	0.022	-0.005
	40	0.859	0.835	0.816	0.060	0.014	0.906	0.033	-0.006
	120	0.617	0.563	0.548	0.075	0.012	0.738	0.040	-0.003
	200	0.497	0.433	0.423	0.074	0.012	0.648	0.039	-0.002
	300	0.409	0.342	0.333	0.071	0.011	0.579	0.038	-0.001
	400	0.353	0.286	0.278	0.068	0.011	0.532	0.036	-0.000

Table A9. Large-deflection Reduction Coefficients K_r

LOADING : UDL intensity q

$W = K_\ell K_r Q$, $\sigma = (t/a)^2 \cdot E\bar{\sigma}$ $W = w/t$

$\bar{\sigma} = K_\ell K_r Q$, $Q = a^4 q / t^4 E$

b/a	Q	W_1	$\bar{\sigma}_{bx1}$	$\bar{\sigma}_{by1}$	$\bar{\sigma}_{mx1}$	$\bar{\sigma}_{my1}$	$\bar{\sigma}_{bx2}$	$\bar{\sigma}_{mx2}$	$\bar{\sigma}_{my2}$
1	Linear K_ℓ ▶	1.633 ×10⁻²	1.47 ×10⁻¹	1.47 ×10⁻¹	1.47 ×10⁻¹	1.47 ×10⁻¹	-2.77 ×10⁻¹	2.77 ×10⁻¹	0.830 ×10⁻¹
	20	0.942	0.928	0.928	0.111	0.111	0.965	0.040	0.048
	40	0.833	0.792	0.792	0.177	0.177	0.894	0.066	0.085
	120	0.541	0.451	0.451	0.242	0.242	0.672	0.100	0.158
	200	0.410	0.314	0.314	0.243	0.243	0.551	0.105	0.184
	300	0.322	0.229	0.229	0.234	0.234	0.460	0.105	0.199
	400	0.269	0.183	0.183	0.225	0.225	0.400	0.102	0.205
1.5	Linear K_ℓ ▶	2.847 ×10⁻²	2.37 ×10⁻¹	1.31 ×10⁻¹	2.37 ×10⁻¹	1.31 ×10⁻¹	-4.26 ×10⁻¹	4.26 ×10⁻¹	1.28 ×10⁻¹
	20	0.868	0.847	0.821	0.119	0.224	0.915	0.059	0.080
	40	0.703	0.659	0.614	0.161	0.307	0.795	0.083	0.126
	120	0.410	0.339	0.300	0.178	0.350	0.546	0.096	0.183
	200	0.303	0.233	0.205	0.170	0.337	0.437	0.092	0.193
	300	0.236	0.171	0.151	0.159	0.320	0.361	0.087	0.194
	400	0.197	0.137	0.122	0.151	0.304	0.313	0.082	0.191
2	Linear K_ℓ ▶	3.420 ×10⁻²	2.68 ×10⁻¹	1.05 ×10⁻¹	2.68 ×10⁻¹	1.05 ×10⁻¹	-4.67 ×10⁻¹	4.67 ×10⁻¹	1.40 ×10⁻¹
	20	0.843	0.824	0.796	0.107	0.291	0.892	0.062	0.105
	40	0.672	0.632	0.596	0.140	0.390	0.758	0.082	0.154
	120	0.390	0.328	0.309	0.150	0.445	0.505	0.088	0.198
	200	0.290	0.229	0.219	0.142	0.433	0.399	0.083	0.202
	300	0.227	0.170	0.165	0.132	0.413	0.324	0.077	0.198
	400	0.190	0.137	0.134	0.125	0.395	0.282	0.072	0.192
3	Linear K_ℓ ▶	3.595 ×10⁻²	2.76 ×10⁻¹	0.847 ×10⁻¹	2.76 ×10⁻¹	0.847 ×10⁻¹	-4.77 ×10⁻¹	4.77 ×10⁻¹	1.43 ×10⁻¹
	20	0.849	0.829	0.832	0.098	0.239	0.885	0.060	0.096
	40	0.685	0.643	0.651	0.129	0.324	0.750	0.079	0.132
	120	0.407	0.343	0.354	0.140	0.372	0.502	0.085	0.152
	200	0.305	0.242	0.252	0.132	0.362	0.399	0.080	0.149
	300	0.240	0.181	0.190	0.124	0.345	0.327	0.075	0.142
	400	0.201	0.147	0.155	0.117	0.330	0.283	0.071	0.136

Table A10. Large-deflection Reduction Coefficients K_r

LOADING: Central Patch Loading, $\alpha = u/a$, $\beta = v/a$

$W = K_\ell K_r P$, $\quad \sigma = (t/a)^2 \cdot E\overline{\sigma}$, $\quad \overline{W} = w/t$

$\overline{\sigma} = K_\ell K_r P$, $\quad P = a^2 p/t^4 E$, $\quad \mathbf{b/a = 1}$

$\alpha \times \beta$	P	W_1	$\overline{\sigma}_{bx1}$	$\overline{\sigma}_{by1}$	$\overline{\sigma}_{mx1}$	$\overline{\sigma}_{my1}$
	Linear K_ℓ →	0.1254	1.72	1.72	1.72	1.72
0.1×0.1	10	0.814	0.848	0.848	0.090	0.090
	20	0.644	0.706	0.706	0.112	0.112
	60	0.387	0.481	0.481	0.119	0.119
	100	0.297	0.392	0.392	0.115	0.115
	150	0.238	0.331	0.331	0.110	0.110
	200	0.203	0.291	0.291	0.106	0.106
	Linear K_ℓ →	0.1210	1.32	1.32	1.32	1.32
0.2×0.2	10	0.825	0.822	0.822	0.106	0.106
	20	0.657	0.651	0.651	0.131	0.131
	60	0.395	0.387	0.387	0.133	0.133
	100	0.301	0.294	0.294	0.123	0.123
	150	0.242	0.234	0.234	0.114	0.114
	200	0.206	0.198	0.198	0.107	0.107
	Linear K_ℓ →	0.1126	1.04	1.04	1.04	1.04
0.3×0.3	10	0.839	0.818	0.818	0.110	0.110
	20	0.674	0.636	0.636	0.137	0.137
	60	0.408	0.356	0.356	0.137	0.137
	100	0.312	0.260	0.260	0.125	0.125
	150	0.250	0.202	0.202	0.114	0.114
	200	0.213	0.168	0.168	0.106	0.106
	Linear K_ℓ →	0.1167	1.20	1.12	1.20	1.12
0.2×0.3	10	0.832	0.824	0.815	0.102	0.115
	20	0.665	0.650	0.634	0.126	0.143
	60	0.401	0.383	0.359	0.124	0.145
	100	0.307	0.289	0.264	0.114	0.135
	150	0.246	0.230	0.206	0.104	0.124
	200	0.210	0.195	0.172	0.096	0.117
	Linear K_ℓ →	0.1117	1.10	0.978	1.10	0.978
0.2×0.4	10	0.841	0.829	0.814	0.098	0.121
	20	0.676	0.656	0.628	0.120	0.152
	60	0.409	0.385	0.343	0.117	0.155
	100	0.313	0.291	0.249	0.106	0.144
	150	0.251	0.231	0.191	0.095	0.133
	200	0.214	0.196	0.158	0.088	0.124

Table A11. Large-deflection Reduction Coefficients K_r

LOADING : Central Patch Loading, $\alpha = u/a$, $\beta = v/a$

$W = K_\ell K_r P$, $\quad \sigma = (t/a)^2 . E\overline{\sigma}$ $\quad \overline{W} = w/t$

$\overline{\sigma} = K_\ell K_r P$, $\quad P = a^2 p / t^4 E$ $\quad$ **b/a = 1.5**

α×β	P	$\overline{W}_1$	$\overline{\sigma}_{bx1}$	$\overline{\sigma}_{by1}$	$\overline{\sigma}_{mx1}$	$\overline{\sigma}_{my1}$
	Linear K_ℓ ➡	0.1664	1.92	1.70	1.92	1.70
0.1x0.1	10	0.783	0.822	0.815	0.079	0.113
	20	0.612	0.680	0.669	0.093	0.135
	60	0.368	0.468	0.453	0.095	0.140
	100	0.282	0.386	0.369	0.091	0.135
	150	0.227	0.328	0.311	0.087	0.129
	200	0.194	0.290	0.273	0.083	0.125
	Linear K_ℓ ➡	0.1616	1.51	1.29	1.51	1.29
0.2x0.2	10	0.793	0.793	0.779	0.089	0.138
	20	0.623	0.626	0.602	0.103	0.164
	60	0.374	0.383	0.353	0.097	0.163
	100	0.286	0.297	0.268	0.088	0.153
	150	0.230	0.240	0.213	0.080	0.143
	200	0.196	0.206	0.180	0.075	0.135
	Linear K_ℓ ➡	0.1528	1.22	1.01	1.22	1.01
0.3x0.3	10	0.804	0.786	0.766	0.088	0.151
	20	0.636	0.607	0.577	0.101	0.180
	60	0.383	0.351	0.313	0.091	0.177
	100	0.294	0.264	0.233	0.081	0.165
	150	0.236	0.209	0.181	0.072	0.152
	200	0.202	0.176	0.151	0.066	0.143
	Linear K_ℓ ➡	0.1577	1.39	1.09	1.39	1.09
0.2x0.3	10	0.799	0.794	0.767	0.083	0.152
	20	0.630	0.625	0.580	0.095	0.182
	60	0.379	0.381	0.326	0.084	0.181
	100	0.291	0.296	0.243	0.075	0.170
	150	0.234	0.240	0.191	0.067	0.158
	200	0.199	0.207	0.161	0.061	0.149
	Linear K_ℓ ➡	0.1532	1.29	0.953	1.29	0.953
0.2x0.4	10	0.806	0.798	0.763	0.077	0.164
	20	0.638	0.628	0.570	0.086	0.197
	60	0.385	0.384	0.313	0.074	0.197
	100	0.296	0.299	0.233	0.064	0.184
	150	0.238	0.244	0.183	0.056	0.171
	200	0.203	0.211	0.154	0.051	0.162

Appendix 161

Table A12. Large-deflection Reduction Coefficients K_r

LOADING : Central Patch Loading, $\alpha = u/a$, $\beta = v/a$

$W = K_\ell K_r P$, $\quad \sigma = (t/a)^2 . E\bar{\sigma}$, $\quad W = w/t$

$\bar{\sigma} = K_\ell K_r P$, $\quad P = a^2 p/t^4 E$, $\quad \mathbf{b/a = 2}$

α×β	P	W_1	$\bar{\sigma}_{bx1}$	$\bar{\sigma}_{by1}$	$\bar{\sigma}_{mx1}$	$\bar{\sigma}_{my1}$
	Linear K_ℓ ➡	0.1795	1.97	1.67	1.97	1.67
0.1×0.1	10	0.801	0.830	0.812	0.077	0.114
	20	0.643	0.695	0.665	0.090	0.137
	60	0.404	0.489	0.449	0.089	0.146
	100	0.314	0.405	0.366	0.085	0.141
	150	0.254	0.345	0.307	0.081	0.136
	200	0.217	0.306	0.269	0.077	0.131
	Linear K_ℓ ➡	0.1746	1.56	1.26	1.56	1.26
0.2×0.2	10	0.810	0.804	0.774	0.085	0.139
	20	0.652	0.646	0.596	0.097	0.167
	60	0.410	0.412	0.350	0.088	0.172
	100	0.318	0.324	0.265	0.079	0.163
	150	0.257	0.265	0.210	0.071	0.153
	200	0.220	0.229	0.177	0.066	0.145
	Linear K_ℓ ➡	0.1657	1.28	0.985	1.28	0.985
0.3×0.3	10	0.822	0.799	0.762	0.083	0.154
	20	0.666	0.632	0.570	0.093	0.185
	60	0.421	0.386	0.314	0.080	0.189
	100	0.327	0.296	0.231	0.069	0.178
	150	0.254	0.238	0.181	0.062	0.167
	200	0.226	0.202	0.151	0.056	0.157
	Linear K_ℓ ➡	0.1708	1.45	1.07	1.45	1.07
0.2×0.3	10	0.817	0.807	0.763	0.079	0.115
	20	0.661	0.648	0.575	0.087	0.186
	60	0.417	0.414	0.326	0.075	0.191
	100	0.324	0.327	0.244	0.065	0.182
	150	0.262	0.269	0.193	0.058	0.171
	200	0.224	0.233	0.163	0.052	0.162
	Linear K_ℓ ➡	0.1665	1.35	0.929	1.35	0.929
0.2×0.4	10	0.826	0.813	0.759	0.071	0.167
	20	0.671	0.654	0.567	0.078	0.203
	60	0.424	0.420	0.317	0.063	0.209
	100	0.330	0.333	0.239	0.054	0.199
	150	0.267	0.275	0.190	0.047	0.187
	200	0.229	0.239	0.161	0.041	0.177

Table A13. Large-deflection Reduction Coefficients K_r

LOADING : Central Patch Loading, $\alpha = u/a$, $\beta = v/a$

$W = K_\ell K_r P$, $\quad (T - (t/a)^2 \cdot EU)$ $\quad W = w/t$

$\bar{\sigma} = K_\ell K_r P$, $\quad P = a^2 p/t^4 E$ $\quad$ **b/a = 3**

$\alpha \times \beta$	P	W_1	$\bar{\sigma}_{bx1}$	$\bar{\sigma}_{by1}$	$\bar{\sigma}_{mx1}$	$\bar{\sigma}_{my1}$
	Linear K_ℓ →	0.1840	1.99	1.66	1.99	1.66
0.1x0.1	10 20 60	0.813 0.671 0.468	0.836 0.709 0.524	0.811 0.663 0.449	0.078 0.091 0.091	0.110 0.131 0.143
	Linear K_ℓ →	0.1791	1.58	1.25	1.58	1.25
0.2x0.2	10 20 60	0.821 0.680 0.474	0.811 0.664 0.456	0.772 0.592 0.349	0.086 0.098 0.088	0.135 0.160 0.167
	Linear K_ℓ →	0.1701	1.30	0.975	1.30	0.975
0.3x0.3	10 20 60	0.834 0.695 0.486	0.808 0.653 0.437	0.760 0.566 0.313	0.084 0.094 0.078	0.148 0.177 0.182
	Linear K_ℓ →	0.1753	1.47	1.06	1.47	1.06
0.2x0.3	10 20 60	0.830 0.690 0.483	0.815 0.668 0.463	0.760 0.571 0.329	0.079 0.088 0.074	0.149 0.177 0.187
	Linear K_ℓ →	0.1711	1.37	0.918	1.37	0.918
0.2x0.4	10 20 60	0.838 0.701 0.493	0.822 0.676 0.473	0.757 0.563 0.326	0.072 0.078 0.062	0.160 0.192 0.204

Appendix 163

Table A14. Large-deflection Reduction Coefficients K_r

LOADING : Central Patch Loading, $\alpha = u/a$, $\beta = v/a$

$W = K_\ell K_r P$, $\quad \sigma = (t/a)^2 \cdot E\bar{\sigma} \quad W = w/t$

$\bar{\sigma} = K_\ell K_r P$, $\quad P = a^2 p/t^4 E \quad$ **b/a = 1**

$\alpha \times \beta$	P	W_1	$\bar{\sigma}_{bx1}$	$\bar{\sigma}_{by1}$	$\bar{\sigma}_{mx1}$	$\bar{\sigma}_{my1}$	$\bar{\sigma}_{bx2}$	$\bar{\sigma}_{mx2}$	$\bar{\sigma}_{my2}$
	Linear K_ℓ ➡	0.06048	1.40	1.40	1.40	1.40	-0.740	0.740	0.222
0.1x0.1	10	0.888	0.916	0.916	0.081	0.081	0.902	0.060	-0.045
	20	0.743	0.805	0.805	0.115	0.115	0.776	0.084	-0.059
	60	0.464	0.574	0.574	0.139	0.139	0.534	0.097	-0.056
	100	0.355	0.473	0.473	0.138	0.138	0.439	0.094	-0.047
	150	0.284	0.400	0.400	0.135	0.135	0.374	0.089	-0.040
	200	0.240	0.352	0.352	0.130	0.130	0.334	0.085	-0.035
	Linear K_ℓ ➡	0.05681	0.999	0.999	0.999	0.999	-0.739	0.739	0.222
0.2x0.2	10	0.895	0.899	0.899	0.099	0.099	0.912	0.055	-0.042
	20	0.752	0.761	0.761	0.140	0.140	0.791	0.078	-0.059
	60	0.469	0.485	0.485	0.163	0.163	0.550	0.089	-0.071
	100	0.358	0.373	0.373	0.158	0.158	0.451	0.085	-0.070
	150	0.284	0.298	0.298	0.149	0.149	0.383	0.079	-0.068
	200	0.240	0.252	0.252	0.142	0.142	0.341	0.075	-0.066
	Linear K_ℓ ➡	0.05050	0.728	0.728	0.728	0.728	-0.708	0.708	0.212
0.3x0.3	10	0.909	0.902	0.902	0.104	0.104	0.926	0.053	-0.027
	20	0.774	0.760	0.760	0.150	0.150	0.817	0.078	-0.039
	60	0.490	0.464	0.464	0.177	0.177	0.580	0.095	-0.044
	100	0.374	0.347	0.347	0.170	0.170	0.479	0.093	-0.041
	150	0.297	0.270	0.270	0.160	0.160	0.409	0.088	-0.037
	200	0.251	0.225	0.225	0.151	0.151	0.366	0.084	-0.034
	Linear K_ℓ ➡	0.05350	0.883	0.811	0.883	0.811	-0.706	0.706	0.212
0.2x0.3	10	0.902	0.903	0.898	0.096	0.108	0.919	0.057	-0.031
	20	0.763	0.767	0.753	0.136	0.154	0.802	0.081	-0.044
	60	0.480	0.486	0.462	0.159	0.182	0.562	0.097	-0.050
	100	0.366	0.373	0.346	0.152	0.176	0.461	0.094	-0.047
	150	0.291	0.297	0.271	0.142	0.166	0.393	0.089	-0.044
	200	0.246	0.252	0.227	0.135	0.158	0.350	0.085	-0.041
	Linear K_ℓ ➡	0.04990	0.786	0.676	0.786	0.676	-0.665	0.665	0.200
0.2x0.4	10	0.911	0.911	0.901	0.092	0.113	0.926	0.057	-0.025
	20	0.778	0.778	0.755	0.133	0.164	0.816	0.083	-0.035
	60	0.494	0.495	0.453	0.155	0.196	0.575	0.102	-0.038
	100	0.377	0.380	0.335	0.148	0.190	0.473	0.100	-0.034
	150	0.300	0.303	0.259	0.138	0.180	0.403	0.096	-0.031
	200	0.254	0.257	0.215	0.130	0.171	0.360	0.091	-0.028

Table A15. Large-deflection Reduction Coefficients K_r

LOADING : Central Patch Loading, $\alpha = u/a$, $\beta = v/a$

$W = K_\ell K_r P$, $\overline{\sigma} = (t/a)^2 \cdot E\overline{\sigma}$, $W = w/t$

$\overline{\sigma} = K_\ell K_r P$, $P = a^2 p/t^4 E$, **b/a = 1.5**

α×β	P	W_1	$\overline{\sigma}_{bx1}$	$\overline{\sigma}_{by1}$	$\overline{\sigma}_{mx1}$	$\overline{\sigma}_{my1}$	$\overline{\sigma}_{bx2}$	$\overline{\sigma}_{mx2}$	$\overline{\sigma}_{my2}$
	Linear K_ℓ →	0.07659	1.54	1.39	1.54	1.39	-0.965	0.965	0.289
0.1x0.1	10	0.850	0.889	0.896	0.094	0.081	0.873	0.073	-0.066
	20	0.689	0.764	0.779	0.125	0.108	0.735	0.094	-0.082
	60	0.417	0.536	0.560	0.143	0.124	0.502	0.100	-0.080
	100	0.317	0.439	0.464	0.141	0.123	0.413	0.094	-0.071
	150	0.251	0.370	0.395	0.136	0.120	0.354	0.087	-0.064
	200	0.213	0.325	0.349	0.132	0.117	0.318	0.082	-0.058
	Linear K_ℓ →	0.07242	1.13	0.984	1.13	0.984	-0.963	0.962	0.289
0.2x0.2	10	0.857	0.867	0.873	0.114	0.098	0.882	0.069	-0.061
	20	0.697	0.715	0.727	0.151	0.131	0.749	0.090	-0.081
	60	0.420	0.446	0.464	0.166	0.144	0.515	0.094	-0.092
	100	0.318	0.341	0.359	0.158	0.138	0.424	0.088	-0.090
	150	0.251	0.272	0.289	0.149	0.130	0.362	0.081	-0.086
	200	0.212	0.230	0.246	0.141	0.124	0.324	0.075	-0.083
	Linear K_ℓ →	0.06583	0.855	0.714	0.855	0.714	-0.926	0.926	0.278
0.3x0.3	10	0.870	0.865	0.869	0.120	0.106	0.896	0.067	-0.046
	20	0.714	0.704	0.713	0.162	0.142	0.771	0.090	-0.062
	60	0.434	0.417	0.431	0.178	0.154	0.538	0.099	-0.066
	100	0.328	0.311	0.324	0.168	0.146	0.446	0.094	-0.062
	150	0.259	0.242	0.255	0.157	0.135	0.383	0.087	-0.057
	200	0.219	0.202	0.213	0.148	0.127	0.344	0.082	-0.054
	Linear K_ℓ →	0.06931	1.01	0.796	1.01	0.796	-0.931	0.931	0.279
0.2x0.3	10	0.862	0.867	0.865	0.111	0.109	0.887	0.071	-0.051
	20	0.703	0.714	0.709	0.147	0.144	0.756	0.094	-0.067
	60	0.425	0.441	0.432	0.160	0.158	0.522	0.101	-0.072
	100	0.321	0.336	0.326	0.151	0.151	0.429	0.095	-0.068
	150	0.254	0.267	0.257	0.141	0.141	0.367	0.088	-0.063
	200	0.214	0.226	0.215	0.133	0.134	0.329	0.083	-0.060
	Linear K_ℓ →	0.06595	0.920	0.660	0.920	0.660	-0.894	0.894	0.268
0.2x0.4	10	0.868	0.871	0.862	0.108	0.116	0.893	0.072	-0.042
	20	0.712	0.717	0.699	0.144	0.155	0.764	0.096	-0.055
	60	0.432	0.442	0.412	0.155	0.169	0.528	0.105	-0.057
	100	0.326	0.337	0.306	0.146	0.160	0.435	0.100	-0.053
	150	0.258	0.268	0.238	0.135	0.150	0.372	0.093	-0.048
	200	0.217	0.227	0.198	0.127	0.142	0.333	0.088	-0.045

Appendix **165**

Table A16. Large-deflection Reduction Coefficients K_r

LOADING : Central Patch Loading, $\alpha = u/a$, $\beta = v/a$

$W = K_\ell K_r P$, $\sigma = (t/a)^2 \cdot E\bar{\sigma}$, $W = w/t$

$\bar{\sigma} = K_\ell K_r P$, $P = a^2 p/t^4 E$,   **b/a = 2**

$\alpha \times \beta$	P	W_1	$\bar{\sigma}_{bx1}$	$\bar{\sigma}_{by1}$	$\bar{\sigma}_{mx1}$	$\bar{\sigma}_{my1}$	$\bar{\sigma}_{bx2}$	$\bar{\sigma}_{mx2}$	$\bar{\sigma}_{my2}$
0.1×0.1	Linear K_ℓ ➡	0.07876	1.55	1.37	1.55	1.37	-0.991	0.991	0.297
	10	0.850	0.888	0.899	0.095	0.075	0.875	0.074	-0.083
	20	0.688	0.764	0.784	0.127	0.100	0.739	0.096	-0.104
	60	0.417	0.535	0.567	0.146	0.117	0.510	0.101	-0.103
	100	0.317	0.438	0.472	0.144	0.116	0.422	0.095	-0.092
	150	0.252	0.369	0.402	0.139	0.114	0.363	0.088	-0.083
	200	0.213	0.324	0.356	0.135	0.111	0.327	0.083	-0.076
0.2×0.2	Linear K_ℓ ➡	0.07445	1.14	0.964	1.14	0.964	-0.987	0.987	0.296
	10	0.856	0.866	0.875	0.115	0.091	0.883	0.071	-0.079
	20	0.695	0.714	0.732	0.152	0.121	0.752	0.092	-0.104
	60	0.420	0.446	0.472	0.168	0.134	0.521	0.096	-0.116
	100	0.317	0.342	0.367	0.161	0.129	0.431	0.089	-0.113
	150	0.251	0.272	0.297	0.151	0.122	0.370	0.082	-0.108
	200	0.212	0.230	0.253	0.144	0.116	0.332	0.076	-0.103
0.3×0.3	Linear K_ℓ ➡	0.06787	0.869	0.698	0.869	0.698	-0.950	0.950	0.285
	10	0.869	0.865	0.871	0.122	0.096	0.897	0.069	-0.060
	20	0.713	0.704	0.717	0.164	0.129	0.774	0.092	-0.081
	60	0.433	0.417	0.437	0.180	0.141	0.545	0.100	-0.088
	100	0.328	0.311	0.331	0.171	0.133	0.453	0.094	-0.084
	150	0.259	0.242	0.261	0.160	0.124	0.391	0.088	-0.078
	200	0.219	0.202	0.219	0.151	0.117	0.352	0.083	-0.073
0.2×0.3	Linear K_ℓ ➡	0.07140	1.03	0.780	1.03	0.780	-0.956	0.956	0.287
	10	0.861	0.867	0.867	0.112	0.099	0.889	0.072	-0.065
	20	0.702	0.713	0.713	0.149	0.132	0.759	0.095	-0.086
	60	0.425	0.440	0.438	0.162	0.146	0.528	0.102	-0.094
	100	0.321	0.336	0.332	0.154	0.139	0.437	0.096	-0.089
	150	0.254	0.267	0.262	0.143	0.131	0.376	0.089	-0.083
	200	0.214	0.225	0.220	0.135	0.124	0.337	0.084	-0.079
0.2×0.4	Linear K_ℓ ➡	0.06813	0.935	0.645	0.935	0.645	-0.920	0.920	0.276
	10	0.867	0.870	0.863	0.109	0.105	0.894	0.073	-0.055
	20	0.710	0.716	0.701	0.145	0.140	0.766	0.098	-0.073
	60	0.430	0.440	0.416	0.157	0.153	0.534	0.106	-0.078
	100	0.325	0.335	0.310	0.148	0.146	0.442	0.101	-0.073
	150	0.257	0.266	0.241	0.137	0.136	0.379	0.094	-0.067
	200	0.217	0.225	0.201	0.129	0.129	0.340	0.088	-0.063

Table A17. Large-deflection Reduction Coefficients K_r

LOADING : Central Patch Loading, $\alpha = u/a$, $\beta = v/a$

$W = K_\ell K_r P$, $\quad \sigma = (t/a)^2 . E\bar{\sigma}$, $\quad \bar{W} = w/t$

$\bar{\sigma} = K_\ell K_r P$, $\quad P = a^2 p/t^4 E$, $\quad$ **b/a = 3**

$\alpha \times \beta$	P	$\bar{W}_1$	$\bar{\sigma}_{bx1}$	$\bar{\sigma}_{by1}$	$\bar{\sigma}_{mx1}$	$\bar{\sigma}_{my1}$	$\bar{\sigma}_{bx2}$	$\bar{\sigma}_{mx2}$	$\bar{\sigma}_{my2}$
	Linear K_ℓ →	0.07887	1.55	1.37	1.55	1.37	-0.991	0.991	0.297
0.1x0.1	10	0.855	0.891	0.902	0.095	0.070	0.881	0.073	-0.104
	20	0.695	0.767	0.788	0.128	0.095	0.749	0.095	-0.133
	60	0.422	0.537	0.571	0.147	0.112	0.522	0.101	-0.135
	100	0.321	0.440	0.476	0.146	0.112	0.434	0.095	-0.123
	150	0.255	0.370	0.406	0.141	0.110	0.375	0.089	-0.112
	200	0.216	0.325	0.360	0.137	0.107	0.338	0.083	-0.104
	Linear K_ℓ →	0.07460	1.14	0.962	1.14	0.962	-0.988	0.988	0.296
0.2x0.2	10	0.860	0.869	0.879	0.114	0.085	0.889	0.070	-0.099
	20	0.701	0.719	0.737	0.153	0.114	0.761	0.091	-0.132
	60	0.425	0.448	0.476	0.170	0.128	0.533	0.096	-0.148
	100	0.322	0.343	0.372	0.163	0.123	0.443	0.089	-0.143
	150	0.255	0.273	0.300	0.154	0.116	0.381	0.082	-0.136
	200	0.215	0.231	0.256	0.146	0.111	0.343	0.077	-0.131
	Linear K_ℓ →	0.06800	0.869	0.696	0.869	0.696	-0.951	0.951	0.285
0.3x0.3	10	0.873	0.868	0.875	0.121	0.089	0.902	0.068	-0.079
	20	0.719	0.709	0.723	0.164	0.120	0.782	0.091	-0.107
	60	0.439	0.421	0.443	0.183	0.132	0.556	0.100	-0.119
	100	0.333	0.313	0.335	0.174	0.124	0.465	0.095	-0.114
	150	0.263	0.244	0.265	0.163	0.116	0.402	0.089	-0.107
	200	0.222	0.203	0.222	0.154	0.109	0.363	0.083	-0.101
	Linear K_ℓ →	0.07153	1.03	0.778	1.03	0.778	-0.957	0.957	0.287
0.2x0.3	10	0.866	0.870	0.871	0.112	0.092	0.894	0.071	-0.085
	20	0.708	0.718	0.718	0.150	0.124	0.768	0.095	-0.114
	60	0.430	0.443	0.442	0.164	0.137	0.539	0.102	-0.125
	100	0.326	0.338	0.336	0.156	0.131	0.449	0.096	-0.119
	150	0.258	0.268	0.265	0.146	0.124	0.387	0.089	-0.111
	200	0.217	0.226	0.223	0.137	0.117	0.348	0.084	-0.106
	Linear K_ℓ →	0.06828	0.936	0.643	0.936	0.643	-0.921	0.921	0.276
0.2x0.4	10	0.871	0.873	0.867	0.109	0.097	0.898	0.072	-0.075
	20	0.716	0.720	0.707	0.146	0.130	0.775	0.097	-0.100
	60	0.436	0.443	0.421	0.159	0.143	0.545	0.107	-0.108
	100	0.330	0.337	0.313	0.150	0.135	0.453	0.101	-0.102
	150	0.261	0.268	0.244	0.139	0.128	0.390	0.094	-0.095
	200	0.220	0.226	0.203	0.131	0.121	0.350	0.089	-0.089

Appendix 167

Table A18. Large-deflection Reduction Coefficients K_r

LOADING : Central Patch Loading, $\alpha = u/a$, $\beta = v/a$

$W = K_\ell K_r P$, $\quad \sigma = (t/a)^2 \cdot E\bar{\sigma} \quad W = w/t$

$\bar{\sigma} = K_\ell K_r P$, $\quad P = a^2 p/t^4 E \quad$ **b/a = 1**

$\alpha \times \beta$	P	W_1	$\bar{\sigma}_{bx1}$	$\bar{\sigma}_{by1}$	$\bar{\sigma}_{mx1}$	$\bar{\sigma}_{my1}$	$\bar{\sigma}_{bx2}$	$\bar{\sigma}_{mx2}$	$\bar{\sigma}_{my2}$
	Linear K_ℓ ➤	0.06048	1.40	1.40	1.40	1.40	-0.740	0.740	0.222
0.1×0.1	10	0.932	0.942	0.942	0.065	0.065	0.949	0.021	-0.201
	20	0.819	0.845	0.845	0.100	0.100	0.863	0.034	-0.305
	60	0.548	0.605	0.605	0.132	0.132	0.655	0.049	-0.389
	100	0.429	0.495	0.495	0.134	0.134	0.560	0.052	-0.384
	150	0.347	0.415	0.415	0.131	0.131	0.493	0.053	-0.367
	200	0.297	0.364	0.364	0.128	0.128	0.449	0.054	-0.350
	Linear K_ℓ ➤	0.05681	0.999	0.999	0.999	0.999	-0.739	0.739	0.222
0.2×0.2	10	0.939	0.933	0.933	0.077	0.077	0.957	0.018	-0.190
	20	0.831	0.816	0.816	0.118	0.118	0.880	0.029	-0.297
	60	0.559	0.528	0.528	0.153	0.153	0.678	0.042	-0.400
	100	0.437	0.403	0.403	0.150	0.150	0.581	0.045	-0.404
	150	0.353	0.320	0.320	0.143	0.143	0.510	0.045	-0.393
	200	0.301	0.269	0.269	0.136	0.136	0.463	0.045	-0.381
	Linear K_ℓ ➤	0.05050	0.728	0.728	0.728	0.728	-0.708	0.708	0.212
0.3×0.3	10	0.951	0.940	0.940	0.075	0.075	0.967	0.019	-0.162
	20	0.856	0.826	0.826	0.120	0.120	0.901	0.032	-0.262
	60	0.590	0.523	0.523	0.160	0.160	0.711	0.049	-0.375
	100	0.464	0.390	0.390	0.158	0.158	0.613	0.052	-0.385
	150	0.375	0.302	0.302	0.149	0.149	0.541	0.054	-0.376
	200	0.321	0.250	0.250	0.141	0.141	0.493	0.054	-0.364
	Linear K_ℓ ➤	0.05350	0.883	0.811	0.883	0.811	-0.706	0.706	0.212
0.2×0.3	10	0.945	0.938	0.934	0.071	0.082	0.961	0.020	-0.175
	20	0.844	0.826	0.815	0.111	0.128	0.889	0.032	-0.278
	60	0.575	0.537	0.513	0.143	0.169	0.691	0.049	-0.383
	100	0.450	0.410	0.383	0.140	0.168	0.592	0.052	-0.388
	150	0.364	0.325	0.297	0.132	0.160	0.520	0.053	-0.378
	200	0.311	0.274	0.247	0.124	0.152	0.473	0.053	-0.365
	Linear K_ℓ ➤	0.04990	0.786	0.675	0.786	0.675	-0.665	0.665	0.200
0.2×0.4	10	0.952	0.945	0.939	0.065	0.083	0.966	0.020	-0.161
	20	0.859	0.840	0.823	0.103	0.133	0.900	0.034	-0.261
	60	0.595	0.554	0.512	0.136	0.182	0.704	0.052	-0.372
	100	0.467	0.425	0.378	0.132	0.182	0.604	0.056	-0.380
	150	0.378	0.338	0.290	0.123	0.174	0.530	0.058	-0.371
	200	0.323	0.286	0.240	0.116	0.167	0.481	0.058	-0.359

Table A19. Large-deflection Reduction Coefficients K_r

LOADING: Central Patch Loading, $\alpha = u/a$, $\beta = v/a$

$W = K_\ell K_r P$, $\quad \sigma = (t/a)^2 . E\bar{\sigma}$ $\quad W = w/t$

$\bar{\sigma} = K_\ell K_r P$, $\quad P = a^2 p/t^4 E$ $\quad$ **b/a = 2**

α×β	P	W_1	$\bar{\sigma}_{bx1}$	$\bar{\sigma}_{by1}$	$\bar{\sigma}_{mx1}$	$\bar{\sigma}_{my1}$	$\bar{\sigma}_{bx2}$	$\bar{\sigma}_{mx2}$	$\bar{\sigma}_{my2}$
	Linear K_ℓ ➡	0.07876	1.55	1.37	1.55	1.37	-0.991	0.991	0.297
0.1×0.1	10	0.892	0.913	0.916	0.083	0.065	0.915	0.050	-0.159
	20	0.749	0.796	0.803	0.117	0.091	0.802	0.071	-0.215
	60	0.476	0.561	0.573	0.143	0.108	0.582	0.084	-0.224
	100	0.368	0.460	0.474	0.142	0.107	0.493	0.082	-0.203
	150	0.297	0.388	0.402	0.139	0.103	0.431	0.077	-0.182
	200	0.254	0.341	0.355	0.135	0.100	0.393	0.074	-0.166
	Linear K_ℓ ➡	0.07445	1.14	0.964	1.14	0.964	-0.987	0.987	0.296
0.2×0.2	10	0.898	0.898	0.898	0.098	0.078	0.923	0.047	-0.151
	20	0.758	0.757	0.757	0.138	0.109	0.814	0.067	-0.209
	60	0.481	0.480	0.481	0.162	0.122	0.594	0.078	-0.234
	100	0.370	0.370	0.370	0.157	0.115	0.501	0.075	-0.221
	150	0.298	0.297	0.296	0.148	0.107	0.436	0.070	-0.204
	200	0.254	0.252	0.251	0.141	0.101	0.395	0.065	-0.192
	Linear K_ℓ ➡	0.06787	0.869	0.698	0.869	0.698	-0.950	0.950	0.285
0.3×0.3	10	0.910	0.901	0.898	0.100	0.079	0.934	0.046	-0.128
	20	0.779	0.756	0.750	0.144	0.112	0.836	0.067	-0.183
	60	0.500	0.462	0.451	0.170	0.124	0.621	0.082	-0.209
	100	0.386	0.347	0.335	0.163	0.115	0.527	0.080	-0.196
	150	0.310	0.272	0.260	0.153	0.105	0.461	0.076	-0.180
	200	0.265	0.228	0.216	0.145	0.097	0.419	0.072	-0.167
	Linear K_ℓ ➡	0.07140	1.03	0.780	1.03	0.780	-0.956	0.956	0.287
0.2×0.3	10	0.904	0.901	0.893	0.093	0.084	0.927	0.049	-0.136
	20	0.767	0.761	0.744	0.132	0.117	0.823	0.070	-0.191
	60	0.489	0.480	0.449	0.153	0.131	0.604	0.084	-0.214
	100	0.378	0.369	0.335	0.146	0.122	0.510	0.081	-0.199
	150	0.304	0.295	0.261	0.136	0.113	0.444	0.076	-0.183
	200	0.259	0.251	0.217	0.128	0.105	0.403	0.072	-0.170
	Linear K_ℓ ➡	0.06813	0.935	0.645	0.935	0.645	-0.920	0.920	0.276
0.2×0.4	10	0.910	0.906	0.892	0.088	0.086	0.932	0.049	-0.125
	20	0.778	0.769	0.737	0.126	0.122	0.831	0.071	-0.177
	60	0.500	0.486	0.430	0.144	0.136	0.611	0.087	-0.199
	100	0.386	0.374	0.314	0.136	0.126	0.516	0.085	-0.184
	150	0.310	0.300	0.242	0.125	0.115	0.450	0.080	-0.168
	200	0.265	0.256	0.200	0.117	0.106	0.407	0.076	-0.156

Appendix 169

Table A20. Large-deflection Reduction Coefficients K_r

LOADING : Central Patch Loading, $\alpha = u/a$, $\beta = v/a$

$W = K_\ell K_r P$, $\quad \sigma = (t/a)^2 . E\bar{\sigma}$ $\quad W = w/t$

$\bar{\sigma} = K_\ell K_r P$, $\quad P = a^2 p/t^4 E$ $\quad \mathbf{b/a = 1}$

αxβ	P	W_1	$\bar{\sigma}_{bx1}$	$\bar{\sigma}_{by1}$	$\bar{\sigma}_{mx1}$	$\bar{\sigma}_{my1}$	$\bar{\sigma}_{bx2}$	$\bar{\sigma}_{by2}$	$\bar{\sigma}_{my2}$
	Linear K_ℓ ➡	0.08574	1.50	1.50	1.50	1.50	-0.409	0.409	0.122
0.1x0.1	10	0.826	0.875	0.875	0.091	0.091	0.830	0.102	-0.116
	20	0.657	0.748	0.748	0.116	0.116	0.666	0.127	-0.145
	60	0.393	0.530	0.530	0.129	0.129	0.413	0.128	-0.150
	100	0.298	0.438	0.438	0.126	0.126	0.322	0.120	-0.141
	150	0.238	0.372	0.372	0.122	0.122	0.264	0.111	-0.131
	200	0.201	0.329	0.329	0.118	0.118	0.228	0.104	-0.124
	Linear K_ℓ ➡	0.08119	1.11	1.11	1.11	1.11	-0.406	0.406	0.122
0.2x0.2	10	0.845	0.859	0.859	0.106	0.106	0.854	0.095	-0.148
	20	0.680	0.705	0.705	0.138	0.138	0.698	0.122	-0.181
	60	0.405	0.441	0.441	0.149	0.149	0.438	0.129	-0.186
	100	0.305	0.339	0.339	0.142	0.142	0.341	0.121	-0.170
	150	0.240	0.271	0.271	0.133	0.133	0.277	0.113	-0.155
	200	0.202	0.230	0.230	0.126	0.126	0.238	0.106	-0.142
	Linear K_ℓ ➡	0.07425	0.838	0.838	0.838	0.838	-0.387	0.387	0.116
0.3x0.3	10	0.875	0.855	0.855	0.113	0.113	0.868	0.092	-0.120
	20	0.696	0.692	0.692	0.148	0.148	0.718	0.122	-0.157
	60	0.418	0.410	0.410	0.158	0.158	0.456	0.131	-0.165
	100	0.315	0.306	0.306	0.148	0.148	0.356	0.124	-0.152
	150	0.249	0.240	0.240	0.138	0.138	0.290	0.116	-0.138
	200	0.209	0.200	0.200	0.129	0.129	0.251	0.109	-0.126
	Linear K_ℓ ➡	0.07756	0.996	0.921	0.996	0.921	-0.392	0.392	0.118
0.2x0.3	10	0.852	0.861	0.853	0.103	0.116	0.861	0.094	-0.133
	20	0.688	0.705	0.690	0.134	0.152	0.706	0.123	-0.171
	60	0.412	0.438	0.412	0.142	0.164	0.444	0.131	-0.177
	100	0.310	0.336	0.309	0.134	0.156	0.346	0.124	-0.162
	150	0.245	0.269	0.243	0.124	0.146	0.281	0.115	-0.147
	200	0.206	0.228	0.203	0.117	0.138	0.242	0.109	-0.135
	Linear K_ℓ ➡	0.07352	0.897	0.780	0.897	0.780	-0.375	0.375	0.112
0.2x0.4	10	0.858	0.865	0.851	0.100	0.124	0.866	0.094	-0.116
	20	0.696	0.710	0.683	0.130	0.163	0.713	0.124	-0.151
	60	0.418	0.441	0.398	0.137	0.177	0.448	0.134	-0.112
	100	0.315	0.339	0.294	0.128	0.168	0.348	0.127	-0.144
	150	0.249	0.272	0.228	0.118	0.157	0.283	0.118	-0.131
	200	0.210	0.231	0.190	0.110	0.148	0.243	0.112	-0.120

Thin Plate Design For Transverse Loading

Table A21. Large-deflection Reduction Coefficients K_r

LOADING : Central Patch Loading, $\alpha = u/a$, $\beta = v/a$

$W = K_\ell K_r P$, $\sigma = (t/a)^2 . E\bar{\sigma}$ $W = w/t$

$\bar{\sigma} = K_\ell K_r P$, $P = a^2 p/t^4 E$ **b/a = 2**

$\alpha \times \beta$	P	W_1	$\bar{\sigma}_{bx1}$	$\bar{\sigma}_{by1}$	$\bar{\sigma}_{mx1}$	$\bar{\sigma}_{my1}$	$\bar{\sigma}_{bx2}$	$\bar{\sigma}_{mx2}$	$\bar{\sigma}_{my2}$
	Linear K_ℓ →	0.1199	1.68	1.47	1.68	1.47	-0.530	0.530	0.159
0.1x0.1	10	0.755	0.828	0.839	0.101	0.089	0.772	0.129	-0.073
	20	0.578	0.695	0.713	0.120	0.107	0.606	0.146	-0.085
	60	0.335	0.489	0.511	0.127	0.112	0.379	0.136	-0.083
	100	0.254	0.404	0.427	0.124	0.109	0.300	0.124	-0.078
	150	0.202	0.344	0.366	0.120	0.104	0.248	0.113	-0.074
	200	0.171	0.304	0.326	0.116	0.101	0.216	0.106	-0.071
	Linear K_ℓ →	0.1144	1.29	1.09	1.29	1.09	-0.524	0.524	0.157
0.2x0.2	10	0.770	0.804	0.810	0.118	0.104	0.793	0.127	-0.090
	20	0.593	0.644	0.655	0.142	0.125	0.631	0.147	-0.105
	60	0.342	0.401	0.412	0.144	0.127	0.397	0.138	-0.096
	100	0.256	0.310	0.319	0.136	0.120	0.312	0.126	-0.084
	150	0.201	0.249	0.257	0.127	0.112	0.256	0.115	-0.072
	200	0.169	0.211	0.218	0.121	0.107	0.222	0.107	-0.065
	Linear K_ℓ →	0.1072	1.02	0.815	1.02	0.815	-0.505	0.505	0.151
0.3x0.3	10	0.780	0.791	0.794	0.125	0.114	0.806	0.125	-0.078
	20	0.603	0.620	0.623	0.149	0.135	0.647	0.143	-0.091
	60	0.348	0.366	0.368	0.148	0.133	0.410	0.140	-0.085
	100	0.260	0.275	0.277	0.138	0.123	0.323	0.128	-0.076
	150	0.205	0.217	0.218	0.127	0.113	0.266	0.117	-0.067
	200	0.172	0.182	0.182	0.119	0.106	0.230	0.109	-0.062
	Linear K_ℓ →	0.1110	1.18	0.897	1.18	0.897	-0.512	0.512	0.154
0.2x0.3	10	0.775	0.801	0.795	0.115	0.115	0.799	0.128	-0.086
	20	0.597	0.638	0.627	0.137	0.137	0.637	0.148	-0.099
	60	0.344	0.394	0.375	0.136	0.137	0.401	0.141	-0.090
	100	0.258	0.304	0.283	0.126	0.128	0.316	0.129	-0.079
	150	0.203	0.244	0.223	0.116	0.119	0.260	0.118	-0.070
	200	0.171	0.208	0.187	0.109	0.112	0.225	0.110	-0.063
	Linear K_ℓ →	0.1069	1.08	0.756	1.08	0.756	-0.494	0.494	0.148
0.2x0.4	10	0.780	0.803	0.787	0.111	0.123	0.805	0.127	-0.073
	20	0.603	0.639	0.612	0.132	0.146	0.645	0.150	-0.086
	60	0.349	0.394	0.354	0.128	0.144	0.408	0.143	-0.081
	100	0.262	0.304	0.263	0.118	0.133	0.322	0.131	-0.072
	150	0.206	0.245	0.205	0.108	0.123	0.262	0.120	-0.065
	200	0.174	0.209	0.171	0.101	0.115	0.230	0.113	-0.060

Table A22. Large-deflection Reduction Coefficients K_r

LOADING : UDL   intensity q

$W = K_\ell K_r Q$, $\sigma = (t/a)^2 . E\bar{\sigma}$ $W = w/t$

$\bar{\sigma} = K_\ell K_r Q$, $Q = a^4 q/t^4 E$

b/a	Q	W_1	$\bar{\sigma}_{bx1}$	$\bar{\sigma}_{by1}$	$\bar{\sigma}_{mx1}$	$\bar{\sigma}_{my1}$	$\bar{\sigma}_{by5}$	$\bar{\sigma}_{mx5}$	$\bar{\sigma}_{my5}$
1	Linear K_ℓ	0.03100	0.206	0.235	0.206	0.235	-0.486	0.146	0.486
	20	0.942	0.913	0.919	0.088	0.086	0.964	-0.081	0.020
	40	0.836	0.759	0.775	0.132	0.129	0.896	-0.126	0.032
	120	0.570	0.414	0.444	0.156	0.154	0.711	-0.165	0.047
	200	0.450	0.283	0.315	0.145	0.145	0.615	-0.162	0.051
	300	0.368	0.205	0.236	0.131	0.134	0.541	-0.151	0.052
	400	0.317	0.161	0.191	0.119	0.124	0.491	-0.140	0.051
1.5	Linear K_ℓ	0.07086	0.415	0.286	0.415	0.286	-0.652	0.196	0.652
	20	0.836	0.787	0.755	0.062	0.203	0.931	-0.115	0.009
	40	0.675	0.591	0.545	0.066	0.249	0.851	-0.159	0.018
	120	0.418	0.306	0.279	0.048	0.249	0.670	-0.188	0.035
	200	0.324	0.211	0.200	0.040	0.230	0.578	-0.180	0.041
	300	0.262	0.152	0.152	0.034	0.210	0.506	-0.165	0.043
	400	0.224	0.118	0.124	0.031	0.194	0.457	-0.152	0.044
2	Linear K_ℓ	0.10180	0.565	0.280	0.565	0.280	-0.692	0.208	0.692
	20	0.828	0.795	0.753	0.023	0.269	0.958	-0.108	0.003
	40	0.670	0.617	0.576	0.021	0.329	0.891	-0.162	0.006
	120	0.418	0.343	0.325	0.016	0.335	0.709	-0.205	0.031
	200	0.323	0.243	0.236	0.014	0.312	0.606	-0.195	0.040
	300	0.261	0.178	0.179	0.012	0.287	0.523	-0.177	0.045
	400	0.223	0.140	0.145	0.011	0.268	0.465	-0.160	0.046

Table A23. Large-deflection Reduction Coefficients K_r

LOADING : UDL   intensity q

$W = K_\ell K_r Q$, $\sigma = (t/a)^2 . E\bar{\sigma}$ $W = w/t$

$\bar{\sigma} = K_\ell K_r Q$, $Q = a^4 q/t^4 E$

b/a	Q	W_1	$\bar{\sigma}_{bx1}$	$\bar{\sigma}_{by1}$	$\bar{\sigma}_{mx1}$	$\bar{\sigma}_{my1}$	$\bar{\sigma}_{bx4}$	$\bar{\sigma}_{mx4}$	$\bar{\sigma}_{my4}$
1.5	Linear K_ℓ ➡	0.04894	0.330	0.177	0.330	0.177	-0.639	0.639	0.192
	20	0.905	0.876	0.842	0.061	0.165	0.928	0.037	-0.056
	40	0.771	0.711	0.650	0.077	0.225	0.825	0.052	-0.076
	120	0.501	0.405	0.353	0.073	0.243	0.602	0.060	-0.075
	200	0.391	0.291	0.255	0.067	0.225	0.500	0.058	-0.063
	300	0.317	0.219	0.195	0.063	0.203	0.426	0.054	-0.051
	400	0.271	0.177	0.160	0.059	0.187	0.378	0.051	-0.042
2	Linear K_ℓ ➡	0.05650	0.368	0.146	0.368	0.146	-0.705	0.705	0.212
	20	0.926	0.909	0.894	0.028	0.175	0.943	0.029	-0.005
	40	0.810	0.771	0.755	0.038	0.253	0.853	0.041	-0.011
	120	0.546	0.469	0.469	0.046	0.304	0.638	0.049	-0.021
	200	0.430	0.342	0.347	0.045	0.293	0.533	0.047	-0.027
	300	0.349	0.258	0.266	0.043	0.274	0.455	0.044	-0.031
	400	0.299	0.208	0.218	0.042	0.257	0.404	0.042	-0.034

Appendix 173

Table A24. Large-deflection Reduction Coefficients K_r

LOADING : UDL   intensity q

$W = K_\ell K_r Q$, $\sigma = (t/a)^2 \cdot E\bar{\sigma}$ $W = w/t$

$\bar{\sigma} = K_\ell K_r Q$, $Q = a^4 q/t^4 E$

b/a	Q	W_1	$\bar{\sigma}_{bx1}$	$\bar{\sigma}_{by1}$	$\bar{\sigma}_{mx1}$	$\bar{\sigma}_{my1}$	$\bar{\sigma}_{bx4}$	$\bar{\sigma}_{mx4}$	$\bar{\sigma}_{my4}$
1	Linear K_ℓ	0.02449	0.185	0.185	0.185	0.185	-0.375	0.375	0.113
	20	0.960	0.942	0.942	0.076	0.076	0.975	0.019	-0.072
	40	0.875	0.822	0.822	0.122	0.122	0.921	0.033	-0.118
	120	0.622	0.495	0.495	0.162	0.162	0.743	0.053	-0.165
	200	0.497	0.353	0.353	0.157	0.157	0.640	0.058	-0.161
	300	0.408	0.264	0.264	0.146	0.146	0.558	0.059	-0.147
	400	0.352	0.214	0.214	0.137	0.137	0.500	0.059	-0.134
1.5	Linear K_ℓ	0.04411	0.302	0.180	0.302	0.180	-0.588	0.588	0.176
	20	0.910	0.881	0.856	0.067	0.153	0.932	0.036	-0.058
	40	0.781	0.718	0.672	0.087	0.214	0.833	0.052	-0.077
	120	0.512	0.407	0.364	0.085	0.244	0.611	0.062	-0.074
	200	0.401	0.290	0.263	0.077	0.232	0.509	0.060	-0.060
	300	0.325	0.216	0.201	0.070	0.216	0.434	0.057	-0.045
	400	0.279	0.173	0.166	0.066	0.203	0.385	0.053	-0.034
2	Linear K_ℓ	0.05421	0.355	0.152	0.355	0.152	-0.683	0.683	0.205
	20	0.920	0.900	0.877	0.037	0.172	0.939	0.031	-0.007
	40	0.802	0.757	0.724	0.047	0.246	0.847	0.044	-0.004
	120	0.544	0.461	0.445	0.050	0.301	0.634	0.051	0.012
	200	0.431	0.337	0.331	0.048	0.298	0.532	0.049	0.021
	300	0.352	0.254	0.256	0.045	0.286	0.456	0.046	0.028
	400	0.303	0.204	0.211	0.043	0.274	0.405	0.044	0.034

174 Thin Plate Design For Transverse Loading

Table A25. Large-deflection Reduction Coefficients K_r

LOADING : UDL   intensity q

$W = K_\ell K_r Q$, $\sigma = (t/a)^2 \cdot E\bar{\sigma}$ $W = w/t$

$\bar{\sigma} = K_\ell K_r Q$, $Q = a^4 q / t^4 E$

b/a	Q	W_1	$\bar{\sigma}_{bx1}$	$\bar{\sigma}_{by1}$	$\bar{\sigma}_{mx1}$	$\bar{\sigma}_{my1}$	$\bar{\sigma}_{by3}$	$\bar{\sigma}_{mx3}$	$\bar{\sigma}_{my3}$
1	Linear K_ℓ	0.02089	0.145	0.197	0.145	0.197	-0.420	0.126	0.420
	20	0.972	0.954	0.961	0.067	0.059	0.984	-0.064	0.015
	40	0.908	0.852	0.873	0.114	0.101	0.945	-0.111	0.026
	120	0.676	0.517	0.575	0.166	0.148	0.800	-0.175	0.045
	200	0.546	0.359	0.426	0.162	0.146	0.711	-0.181	0.051
	300	0.451	0.258	0.327	0.150	0.137	0.640	-0.175	0.052
	400	0.390	0.202	0.269	0.138	0.128	0.592	-0.166	0.052
1.5	Linear K_ℓ	0.05803	0.348	0.274	0.348	0.274	-0.630	0.189	0.630
	20	0.868	0.820	0.796	0.077	0.171	0.939	-0.127	0.008
	40	0.717	0.626	0.587	0.089	0.222	0.863	-0.178	0.015
	120	0.453	0.324	0.295	0.070	0.233	0.694	-0.216	0.030
	200	0.351	0.221	0.208	0.056	0.217	0.610	-0.209	0.036
	300	0.284	0.158	0.157	0.047	0.200	0.545	-0.195	0.039
	400	0.243	0.122	0.129	0.042	0.185	0.500	-0.181	0.041
2	Linear K_ℓ	0.09222	0.519	0.284	0.519	0.284	-0.717	0.215	0.717
	20	0.828	0.787	0.735	0.035	0.250	0.948	-0.134	-0.007
	40	0.671	0.607	0.548	0.032	0.305	0.879	-0.196	-0.001
	120	0.419	0.335	0.305	0.022	0.310	0.714	-0.246	0.020
	200	0.325	0.237	0.221	0.018	0.289	0.627	-0.239	0.030
	300	0.262	0.173	0.168	0.016	0.267	0.560	-0.221	0.035
	400	0.224	0.136	0.138	0.017	0.250	0.513	-0.204	0.037

Table A26. Large-deflection Reduction Coefficients K_r

LOADING : UDL intensity q

$W = K_\ell K_r Q$, $\sigma = (t/a)^2 \cdot E\bar{\sigma}$ $W = w/t$

$\bar{\sigma} = K_\ell K_r Q$, $Q = a^4 q / t^4 E$

b/a	Q	W_1	$\bar{\sigma}_{bx1}$	$\bar{\sigma}_{by1}$	$\bar{\sigma}_{mx1}$	$\bar{\sigma}_{my1}$	$\bar{\sigma}_{bx2}$	$\bar{\sigma}_{mx2}$	$\bar{\sigma}_{my2}$
1.5	Linear K_ℓ ➡	0.02706	0.240	0.106	0.240	0.106	-0.495	0.495	0.148
	20	0.957	0.945	0.919	0.052	0.083	0.973	0.025	-0.052
	40	0.872	0.839	0.772	0.079	0.128	0.918	0.041	-0.083
	120	0.621	0.549	0.438	0.092	0.157	0.749	0.057	-0.105
	200	0.497	0.418	0.318	0.086	0.146	0.658	0.057	-0.099
	300	0.408	0.329	0.246	0.079	0.131	0.588	0.054	-0.090
	400	0.353	0.276	0.205	0.073	0.119	0.540	0.052	-0.082
2	Linear K_ℓ ➡	0.02852	0.250	0.0848	0.250	0.0848	-0.507	0.507	0.152
	20	0.955	0.945	0.918	0.044	0.048	0.971	0.025	-0.024
	40	0.864	0.837	0.775	0.064	0.067	0.912	0.039	-0.035
	120	0.621	0.564	0.498	0.075	0.071	0.747	0.049	-0.038
	200	0.501	0.435	0.384	0.072	0.064	0.660	0.047	-0.035
	300	0.414	0.346	0.305	0.068	0.057	0.593	0.044	-0.031
	400	0.359	0.291	0.257	0.065	0.052	0.548	0.042	-0.028

Table A27. Large-deflection Reduction Coefficients K_r

LOADING : UDL intensity q

$W = K_\ell K_r Q,$ $\sigma = (t/a)^2 . E\bar{\sigma}$ $W = w/t$

$\bar{\sigma} = K_\ell K_r Q,$ $Q = a^4 q/t^4 E$

b/a	Q	W_1	$\bar{\sigma}_{bx1}$	$\bar{\sigma}_{by1}$	$\bar{\sigma}_{mx1}$	$\bar{\sigma}_{my1}$	$\bar{\sigma}_{by3}$	$\bar{\sigma}_{mx3}$	$\bar{\sigma}_{my3}$
1	Linear K_ℓ	0.02089	0.145	0.197	0.145	0.197	-0.420	0.126	0.420
	20	0.945	0.923	0.929	0.063	0.104	0.967	-0.060	0.037
	40	0.842	0.782	0.798	0.097	0.163	0.905	-0.095	0.059
	120	0.565	0.436	0.473	0.114	0.214	0.723	-0.123	0.084
	200	0.437	0.294	0.337	0.104	0.210	0.627	-0.117	0.086
	300	0.349	0.207	0.252	0.091	0.198	0.554	-0.108	0.085
	400	0.296	0.159	0.204	0.082	0.187	0.504	-0.099	0.082
1.5	Linear K_ℓ	0.05803	0.348	0.274	0.348	0.274	-0.630	0.189	0.630
	20	0.814	0.768	0.743	0.067	0.238	0.911	-0.112	0.045
	40	0.642	0.565	0.531	0.072	0.291	0.817	-0.143	0.061
	120	0.377	0.278	0.261	0.055	0.289	0.629	-0.149	0.073
	200	0.283	0.186	0.181	0.045	0.266	0.541	-0.137	0.074
	300	0.222	0.131	0.135	0.038	0.243	0.475	-0.123	0.072
	400	0.187	0.101	0.109	0.034	0.226	0.430	-0.112	0.070
2	Linear K_ℓ	0.09222	0.519	0.284	0.519	0.284	-0.717	0.215	0.717
	20	0.774	0.737	0.691	0.032	0.332	0.923	-0.117	0.039
	40	0.600	0.545	0.499	0.029	0.387	0.837	-0.154	0.054
	120	0.349	0.282	0.261	0.021	0.374	0.652	-0.165	0.069
	200	0.262	0.193	0.185	0.018	0.343	0.561	-0.152	0.070
	300	0.206	0.139	0.138	0.016	0.313	0.491	-0.137	0.070
	400	0.173	0.108	0.111	0.014	0.291	0.444	-0.125	0.068

Appendix 177

Table A28. Large-deflection Reduction Coefficients K_r

LOADING : UDL   intensity   q

$W = K_\ell K_r Q$, $\sigma = (t/a)^2 . E\bar{\sigma}$ $W = w/t$

$\bar{\sigma} = K_\ell K_r Q$, $Q = a^4 q / t^4 F$

b/a	Q	W_1	$\bar{\sigma}_{bx1}$	$\bar{\sigma}_{by1}$	$\bar{\sigma}_{mx1}$	$\bar{\sigma}_{my1}$	$\bar{\sigma}_{bx2}$	$\bar{\sigma}_{mx2}$	$\bar{\sigma}_{my2}$
1.5	Linear K_ℓ ▶	0.02706	0.240	0.106	0.240	0.106	-0.495	0.495	0.148
	20	0.901	0.880	0.851	0.109	0.072	0.936	0.053	-0.046
	40	0.759	0.711	0.652	0.153	0.094	0.840	0.076	-0.062
	120	0.468	0.390	0.311	0.169	0.087	0.627	0.089	-0.059
	200	0.353	0.275	0.205	0.159	0.072	0.532	0.086	-0.051
	300	0.279	0.205	0.147	0.148	0.061	0.463	0.080	-0.043
	400	0.234	0.166	0.116	0.139	0.053	0.418	0.076	-0.038
2	Linear K_ℓ ▶	0.02852	0.250	0.0848	0.250	0.0848	-0.507	0.507	0.152
	20	0.879	0.856	0.825	0.113	0.039	0.919	0.059	-0.020
	40	0.719	0.672	0.618	0.150	0.044	0.810	0.079	-0.023
	120	0.433	0.363	0.314	0.163	0.033	0.598	0.086	-0.018
	200	0.326	0.256	0.219	0.154	0.026	0.509	0.081	-0.014
	300	0.257	0.191	0.164	0.144	0.021	0.446	0.076	-0.011
	400	0.216	0.155	0.133	0.136	0.018	0.405	0.071	-0.010

Thin Plate Design For Transverse Loading

Table A29. Large-deflection Reduction Coefficients K_r

LOADING : UDL   intensity q

$W = K_\ell K_r Q$, $\sigma = (t/a)^2 . E\bar{\sigma}$ $\bar{W} = w/t$

$\bar{\sigma} = K_\ell K_r Q$, $Q = a^4 q / t^4 E$

b/a	Q	W_1	W_3	$\bar{\sigma}_{bx1}$	$\bar{\sigma}_{by1}$	$\bar{\sigma}_{mx1}$	$\bar{\sigma}_{my1}$
1	Linear K_ℓ ➤	0.06007	0.1127	0.330	0.174	0.330	0.174
	20	1.000	1.004	0.986	0.985	0.031	0.006
	40	1.031	1.045	0.974	0.969	0.058	-0.000
	60	1.214	1.235	1.046	1.044	0.096	-0.019
1.5	Linear K_ℓ ➤	0.1104	0.1509	0.599	0.255	0.599	0.255
	20	1.01	1.20	0.986	0.814	0.019	0.077
2	Linear K_ℓ ➤	0.1156	0.1517	0.624	0.253	0.624	0.253
	20	1.02	1.36	0.993	0.810	0.014	0.047

Appendix 179

Table A30. Large-deflection Reduction Coefficients K_r

LOADING : UDL   intensity q

$W = K_\ell K_r Q$, $\sigma = (t/a)^2 . E\bar{\sigma}$ $W = w/t$

$\bar{\sigma} = K_\ell K_r Q$, $Q = a^4 q/t^4 E$

b/a	Q	W_1	W_3	$\bar{\sigma}_{bx1}$	$\bar{\sigma}_{by1}$	$\bar{\sigma}_{mx1}$	$\bar{\sigma}_{my1}$	$\bar{\sigma}_{bx3}$	σ_{mx3}
1	Linear K_ℓ	0.02050	0.02899	0.186	0.103	0.186	0.103	0.236	0.236
	20	0.976	0.950	0.974	0.994	0.018	0.053	0.944	0.069
	40	0.925	0.856	0.917	0.966	0.039	0.099	0.840	0.103
	120	0.711	0.589	0.659	0.717	0.087	0.182	0.548	0.117
	200	0.581	0.467	0.506	0.543	0.096	0.198	0.417	0.111
	300	0.482	0.381	0.395	0.418	0.094	0.198	0.328	0.104
	400	0.417	0.327	0.326	0.345	0.091	0.194	0.274	0.097
1.5	Linear K_ℓ	0.02612	0.02897	0.231	0.0978	0.231	0.0978	0.238	0.238
	20	0.973	0.955	0.967	0.962	0.032	0.059	0.949	0.057
	40	0.909	0.871	0.889	0.864	0.056	0.102	0.854	0.089
	120	0.860	0.635	0.619	0.544	0.078	0.153	0.591	0.123
	200	0.553	0.513	0.479	0.407	0.077	0.157	0.458	0.123
	300	0.456	0.423	0.378	0.315	0.071	0.151	0.363	0.118
	400	0.394	0.365	0.316	0.263	0.068	0.144	0.305	0.112
2	Linear K_ℓ	0.02809	0.02761	0.245	0.0846	0.245	0.0846	0.228	0.228
	20	0.970	0.972	0.962	0.946	0.031	0.041	0.968	0.038
	40	0.902	0.911	0.881	0.835	0.049	0.064	0.898	0.065
	120	0.680	0.700	0.627	0.560	0.061	0.080	0.662	0.106
	200	0.557	0.576	0.493	0.437	0.060	0.081	0.527	0.113
	300	0.464	0.479	0.395	0.350	0.058	0.080	0.424	0.111
	400	0.403	0.417	0.334	0.295	0.056	0.078	0.360	0.108

Table A31. Large-deflection Reduction Coefficients K_r

LOADING : UDL   intensity q

$W = K_\ell K_r Q$, $\sigma = (t/a)^2 \cdot E\bar{\sigma}$, $W = w/t$

$\bar{\sigma} = K_\ell K_r Q$, $Q = a^4 q / t^4 E$

b/a	Q	W_1	W_3	σ_{bx1}	σ_{by1}	σ_{mx1}	σ_{my1}	σ_{bx3}	σ_{mx3}
	Linear K_ℓ ➤	0.02050	0.02899	0.186	0.103	0.186	0.103	0.236	0.236
1	20	0.925	0.892	0.914	0.941	0.097	0.048	0.879	0.123
	40	0.810	0.741	0.781	0.834	0.148	0.075	0.711	0.169
	120	0.541	0.446	0.476	0.538	0.197	0.103	0.392	0.181
	200	0.422	0.335	0.348	0.395	0.197	0.102	0.278	0.170
	300	0.339	0.263	0.264	0.299	0.189	0.097	0.208	0.157
	400	0.288	0.221	0.215	0.242	0.180	0.091	0.169	0.148
	Linear K_ℓ ➤	0.02612	0.02897	0.231	0.0978	0.231	0.0978	0.238	0.238
1.5	20	0.900	0.877	0.881	0.876	0.108	0.050	0.861	0.125
	40	0.760	0.723	0.718	0.700	0.153	0.070	0.687	0.170
	120	0.475	0.440	0.401	0.364	0.175	0.072	0.380	0.188
	200	0.361	0.332	0.285	0.247	0.166	0.064	0.271	0.178
	300	0.286	0.262	0.213	0.179	0.155	0.057	0.203	0.166
	400	0.241	0.221	0.173	0.142	0.147	0.051	0.165	0.156
	Linear K_ℓ ➤	0.02809	0.02761	0.245	0.0846	0.245	0.0846	0.228	0.228
2	20	0.883	0.884	0.862	0.842	0.110	0.033	0.868	0.121
	40	0.729	0.733	0.684	0.647	0.149	0.040	0.699	0.166
	120	0.443	0.452	0.372	0.332	0.164	0.033	0.395	0.187
	200	0.336	0.344	0.265	0.232	0.157	0.027	0.285	0.180
	300	0.265	0.272	0.198	0.172	0.146	0.023	0.215	0.168
	400	0.223	0.229	0.160	0.138	0.138	0.020	0.175	0.159

Appendix 181

Table A32. Large-deflection Reduction Coefficients K_r

LOADING : UDL   intensity q

$W = K_\ell K_r Q$, $\sigma = (t/a)^2 \cdot E\bar{\sigma}$ $\overline{W} = w/t$

$\bar{\sigma} = K_\ell K_r Q$, $Q = a^4 q / t^4 E$

b/a	Q	$\overline{W}_1$	$\bar{\sigma}_{bx1}$	$\bar{\sigma}_{by1}$	$\bar{\sigma}_{mx1}$	$\bar{\sigma}_{my1}$	$\bar{\sigma}_{bx4}$	$\bar{\sigma}_{mx4}$	$\bar{\sigma}_{my4}$
1	Linear K_ℓ	0.02767	0.198	0.198	0.198	0.198	-0.311	0.311	0.0934
	20	0.953	0.932	0.932	0.077	0.077	0.976	0.018	-0.046
	40	0.866	0.807	0.807	0.121	0.121	0.923	0.033	-0.075
	120	0.619	0.486	0.486	0.158	0.158	0.729	0.061	-0.105
	200	0.497	0.349	0.349	0.152	0.152	0.609	0.068	-0.101
	300	0.409	0.261	0.261	0.140	0.140	0.515	0.069	-0.089
	400	0.354	0.210	0.210	0.129	0.129	0.451	0.069	-0.077
1.5	Linear K_ℓ	0.05040	0.324	0.199	0.324	0.199	-0.473	0.473	0.142
	20	0.891	0.860	0.829	0.069	0.152	0.931	0.041	0.027
	40	0.757	0.693	0.639	0.086	0.202	0.827	0.060	0.029
	120	0.501	0.397	0.350	0.081	0.224	0.583	0.070	0.073
	200	0.397	0.285	0.256	0.071	0.215	0.470	0.066	0.084
2	Linear K_ℓ	0.06500	0.396	0.177	0.396	0.177	-0.542	0.542	0.163
	20	0.891	0.871	0.838	0.036	0.177	0.928	0.037	0.022
	40	0.765	0.723	0.681	0.043	0.236	0.825	0.050	0.047
	80	0.612	0.545	0.515	0.043	0.272	0.680	0.056	0.078
	120	0.521	0.441	0.423	0.042	0.282	0.587	0.056	0.091

Thin Plate Design For Transverse Loading

Table A33. Large-deflection Reduction Coefficients K_r

LOADING : UDL   intensity q

$W = K_\ell K_r Q$, $\sigma = (t/a)^2 \cdot E\bar{\sigma}$ $W = w/t$

$\bar{\sigma} = K_\ell K_r Q$, $Q = a^4 q / t^4 E$

b/a	Q	W_1	$\bar{\sigma}_{bx1}$	$\bar{\sigma}_{by1}$	$\bar{\sigma}_{mx1}$	$\bar{\sigma}_{my1}$	$\bar{\sigma}_{bx2}$	$\bar{\sigma}_{mx2}$	$\bar{\sigma}_{my2}$
1	Linear K_ℓ	0.01955	0.171	0.154	0.171	0.154	-0.312	0.312	0.0936
	20	0.957	0.942	0.940	0.096	0.067	0.971	0.036	-0.040
	40	0.871	0.828	0.821	0.158	0.107	0.911	0.060	-0.065
	120	0.606	0.504	0.484	0.223	0.140	0.705	0.094	-0.083
	200	0.475	0.362	0.336	0.224	0.133	0.585	0.100	-0.075
	300	0.383	0.272	0.242	0.214	0.122	0.493	0.100	-0.064
	400	0.325	0.221	0.189	0.203	0.113	0.430	0.097	-0.055
1.5	Linear K_ℓ	0.03204	0.252	0.134	0.252	0.134	-0.422	0.422	0.127
	20	0.879	0.854	0.821	0.120	0.129	0.913	0.064	-0.006
	40	0.722	0.669	0.608	0.162	0.168	0.786	0.090	-0.000
	120	0.431	0.351	0.279	0.172	0.171	0.519	0.103	0.024
	200	0.323	0.247	0.186	0.160	0.159	0.405	0.098	0.035
	300	0.253	0.185	0.135	0.148	0.147	0.328	0.091	0.043
	400	0.213	0.150	0.108	0.138	0.138	0.280	0.085	0.047
2	Linear K_ℓ	0.03610	0.277	0.106	0.277	0.106	-0.464	0.464	0.139
	20	0.851	0.827	0.787	0.110	0.141	0.887	0.065	0.018
	40	0.683	0.635	0.576	0.142	0.179	0.746	0.086	0.033
	120	0.401	0.332	0.288	0.147	0.193	0.480	0.090	0.057
	200	0.301	0.234	0.204	0.137	0.189	0.375	0.084	0.063
	300	0.237	0.176	0.155	0.127	0.181	0.304	0.077	0.065
	400	0.199	0.143	0.127	0.120	0.175	0.260	0.072	0.066

Appendix 183

Table A34. Large-deflection Reduction Coefficients K_r

LOADING : UDL intensity q

$W = K_\ell K_r Q$, $\sigma = (t/a)^2 \cdot E\bar{\sigma}$ $W = w/t$

$\bar{\sigma} = K_\ell K_r Q$, $Q = a^4 q / t^4 E$

b/a	Q	W_1	$\bar{\sigma}_{bx1}$	$\bar{\sigma}_{by1}$	$\bar{\sigma}_{mx1}$	$\bar{\sigma}_{my1}$	$\bar{\sigma}_{bx4}$	$\bar{\sigma}_{mx4}$	$\bar{\sigma}_{my4}$
1.5	Linear K_ℓ ➡	0.04361	0.289	0.194	0.289	0.194	-0.481	0.481	0.144
	20	0.885	0.853	0.829	0.076	0.182	0.919	0.046	0.053
	40	0.741	0.674	0.635	0.097	0.245	0.803	0.065	0.089
	120	0.468	0.363	0.336	0.091	0.275	0.553	0.074	0.149
	200	0.360	0.253	0.240	0.081	0.263	0.444	0.070	0.168
	300	0.288	0.184	0.183	0.073	0.245	0.367	0.065	0.176
2	Linear K_ℓ ➡	0.06044	0.375	0.177	0.375	0.177	-0.559	0.559	0.168
	20	0.879	0.857	0.822	0.041	0.217	0.926	0.036	0.088
	40	0.741	0.697	0.652	0.048	0.288	0.823	0.050	0.139
	120	0.482	0.404	0.384	0.047	0.332	0.588	0.056	0.200
	200	0.375	0.288	0.282	0.044	0.322	0.479	0.054	0.208
	250	0.339	0.247	0.247	0.043	0.309	0.439	0.052	0.209

Table A35. Large-deflection Reduction Coefficients K_r

LOADING : Linearly varying pressure

$W = K_\ell K_r Q$, $\bar{\sigma} = (t/a)^2 \cdot E\bar{\sigma}$, $W = w/t$

$\bar{\sigma} = K_\ell K_r Q$, $Q = a^4 q_{max}/t^4 E$

b/a	Q	W_1	$\bar{\sigma}_{bx1}$	$\bar{\sigma}_{by1}$	$\bar{\sigma}_{mx1}$	$\bar{\sigma}_{my1}$
0.67	Linear K_ℓ ➡	9.216 ×10^{-3}	7.30 ×10^{-2}	1.19 ×10^{-1}	7.30 ×10^{-2}	1.19 ×10^{-1}
	20	0.993	0.989	0.991	0.035	0.023
	40	0.969	0.956	0.965	0.066	0.043
	120	0.810	0.735	0.785	0.135	0.082
	200	0.684	0.577	0.647	0.157	0.089
	300	0.577	0.457	0.534	0.165	0.086
	400	0.503	0.382	0.459	0.165	0.082
1	Linear K_ℓ ➡	2.418 ×10^{-2}	1.56 ×10^{-1}	1.56 ×10^{-1}	1.56 ×10^{-1}	1.56 ×10^{-1}
	20	0.936	0.922	0.922	0.082	0.068
	40	0.822	0.784	0.785	0.125	0.101
	120	0.540	0.465	0.461	0.154	0.118
	200	0.416	0.337	0.330	0.147	0.110
	300	0.332	0.258	0.247	0.136	0.099
	400	0.281	0.212	0.200	0.127	0.092
1.5	Linear K_ℓ ➡	4.482 ×10^{-2}	2.57 ×10^{-1}	1.58 ×10^{-1}	2.57 ×10^{-1}	1.58 ×10^{-1}
	20	0.810	0.789	0.743	0.109	0.084
	40	0.634	0.599	0.519	0.125	0.096
	120	0.368	0.328	0.233	0.107	0.083
	200	0.275	0.240	0.155	0.090	0.072
	300	0.217	0.187	0.114	0.077	0.063
	400	0.183	0.156	0.093	0.069	0.057
2	Linear K_ℓ ➡	5.881 ×10^{-2}	3.23 ×10^{-1}	1.47 ×10^{-1}	3.23 ×10^{-1}	1.47 ×10^{-1}
	20	0.716	0.695	0.606	0.100	0.058
	40	0.539	0.511	0.393	0.100	0.054
	120	0.306	0.280	0.185	0.074	0.038
	200	0.229	0.207	0.135	0.062	0.031
	300	0.181	0.163	0.106	0.053	0.027
	400	0.153	0.137	0.089	0.048	0.024

Appendix 185

Table A36. Large-deflection Reduction Coefficients K_r

LOADING : Linearly varying pressure

$W = K_\ell K_r Q$, $\quad \bar{\sigma} = (t/a)^2 \cdot E\bar{\sigma}$, $\quad W = w/t$

$\bar{\sigma} = K_\ell K_r Q$, $\quad Q = a^4 q_{max}/t^4 E$

b/a	Q	W_1	$\bar{\sigma}_{bx1}$	$\bar{\sigma}_{by1}$	$\bar{\sigma}_{mx1}$	$\bar{\sigma}_{my1}$
0.67	Linear K_ℓ	9.285×10^{-3}	7.38×10^{-2}	1.20×10^{-1}	7.38×10^{-2}	1.20×10^{-1}
	20	0.986	0.982	0.985	0.064	0.023
	40	0.951	0.936	0.946	0.119	0.042
	120	0.760	0.698	0.737	0.229	0.074
	200	0.625	0.544	0.593	0.260	0.078
	300	0.517	0.430	0.480	0.268	0.076
	400	0.445	0.359	0.407	0.266	0.072
1	Linear K_ℓ	2.418×10^{-2}	1.56×10^{-1}	1.56×10^{-1}	1.56×10^{-1}	1.56×10^{-1}
	20	0.885	0.869	0.868	0.149	0.061
	40	0.732	0.700	0.698	0.204	0.082
	120	0.442	0.391	0.383	0.224	0.085
	200	0.331	0.281	0.270	0.209	0.077
	300	0.259	0.214	0.201	0.192	0.070
	400	0.217	0.175	0.161	0.179	0.064
1.5	Linear K_ℓ	4.482×10^{-2}	2.57×10^{-1}	1.58×10^{-1}	2.57×10^{-1}	1.58×10^{-1}
	20	0.681	0.659	0.619	0.176	0.061
	40	0.493	0.463	0.410	0.184	0.062
	120	0.263	0.234	0.181	0.157	0.049
	200	0.190	0.164	0.118	0.136	0.041
	300	0.146	0.124	0.084	0.121	0.035
	400	0.121	0.102	0.067	0.111	0.032
2	Linear K_ℓ	5.881×10^{-2}	3.23×10^{-1}	1.47×10^{-1}	3.23×10^{-1}	1.47×10^{-1}
	20	0.564	0.543	0.482	0.165	0.039
	40	0.390	0.366	0.301	0.158	0.033
	120	0.200	0.179	0.131	0.125	0.022
	200	0.143	0.126	0.089	0.108	0.017
	300	0.111	0.096	0.066	0.097	0.015
	400	0.092	0.079	0.054	0.089	0.013

Table A37. Large-deflection Reduction Coefficients K_r

LOADING: Linearly varying pressure

$W = K_\ell K_r Q$, $\quad \sigma = (t/a)^2 \cdot E\bar{\sigma}$, $\quad W = w/t$

$\bar{\sigma} = K_\ell K_r Q$, $\quad Q = a^4 q_{max}/t^4 E$

b/a	Q	W_1	W_3	$\bar{\sigma}_{bx1}$	$\bar{\sigma}_{by1}$	$\bar{\sigma}_{mx1}$	$\bar{\sigma}_{my1}$	$\bar{\sigma}_{bx3}$	$\bar{\sigma}_{mx3}$
0.67	Linear K_ℓ ➤	1.681×10^{-2}	2.374×10^{-2}	1.04×10^{-1}	8.55×10^{-2}	1.04×10^{-1}	8.55×10^{-2}	1.15×10^{-1}	1.15×10^{-1}
	20	0.952	0.934	0.956	0.979	0.013	0.050	0.935	0.071
	40	0.860	0.812	0.870	0.932	0.027	0.083	0.816	0.095
	120	0.603	0.508	0.607	0.734	0.059	0.130	0.516	0.058
	200	0.476	0.381	0.468	0.599	0.068	0.136	0.390	0.025
	300	0.384	0.296	0.368	0.492	0.070	0.133	0.307	0.000
	400	0.328	0.247	0.306	0.422	0.070	0.126	0.258	-0.013
1	Linear K_ℓ ➤	2.401×10^{-2}	2.315×10^{-2}	1.43×10^{-1}	9.97×10^{-2}	1.43×10^{-1}	9.97×10^{-2}	1.13×10^{-1}	1.13×10^{-1}
	20	0.941	0.917	0.938	0.945	0.048	0.063	0.918	0.017
	40	0.830	0.780	0.818	0.829	0.076	0.100	0.782	0.008
	120	0.550	0.500	0.516	0.517	0.098	0.131	0.505	-0.018
	200	0.425	0.387	0.385	0.380	0.094	0.126	0.395	-0.021
	300	0.340	0.313	0.299	0.291	0.086	0.116	0.323	-0.018
	400	0.288	0.268	0.250	0.240	0.080	0.108	0.279	-0.015
1.5	Linear K_ℓ ➤	3.456×10^{-2}	1.820×10^{-2}	1.95×10^{-1}	1.03×10^{-1}	1.95×10^{-1}	1.03×10^{-1}	8.92×10^{-2}	8.92×10^{-2}
	20	0.884	1.03	0.869	0.815	0.078	0.071	1.034	-0.139
	40	0.743	1.02	0.714	0.612	0.095	0.088	1.022	-0.115
	80	0.567	0.890	0.531	0.408	0.091	0.089	0.892	-0.036
2	Linear K_ℓ ➤	4.218×10^{-2}	1.235×10^{-2}	2.30×10^{-1}	9.70×10^{-2}	2.30×10^{-1}	9.70×10^{-2}	6.06×10^{-2}	6.06×10^{-2}
	20	0.813	1.446	0.797	0.718	0.073	0.056	1.446	-0.386
	40	0.674	1.739	0.649	0.518	0.076	0.051	0.355	-0.222

Appendix 187

Table A38. Large-deflection Reduction Coefficients K_r

LOADING : Linearly varying pressure

$W = K_\ell K_r Q$, $\quad \bar{\sigma} = (t/a)^2 \cdot E\bar{\sigma}$ $\quad W = w/t$

$\bar{\sigma} = K_\ell K_r Q$, $\quad Q = a^4 q_{max}/t^4 E$

b/a	Q	W_1	W_3	$\bar{\sigma}_{bx1}$	$\bar{\sigma}_{by1}$	$\bar{\sigma}_{mx1}$	$\bar{\sigma}_{my1}$	$\bar{\sigma}_{bx3}$	$\bar{\sigma}_{mx3}$
0.67	Linear K_ℓ ➤	1.681×10^{-2}	2.374×10^{-2}	1.04×10^{-1}	8.55×10^{-2}	1.04×10^{-1}	8.55×10^{-2}	1.15×10^{-1}	1.15×10^{-1}
	20	0.882	0.846	0.890	0.935	0.109	0.044	0.846	0.141
	40	0.739	0.666	0.754	0.845	0.153	0.064	0.666	0.169
	120	0.470	0.356	0.486	0.629	0.187	0.086	0.357	0.124
	200	0.362	0.248	0.371	0.515	0.186	0.089	0.249	0.086
	300	0.289	0.182	0.292	0.428	0.178	0.088	0.183	0.057
	400	0.244	0.145	0.244	0.370	0.171	0.085	0.145	0.039
1	Linear K_ℓ ➤	2.401×10^{-2}	2.315×10^{-2}	1.43×10^{-1}	9.97×10^{-2}	1.43×10^{-1}	9.97×10^{-2}	1.13×10^{-1}	1.13×10^{-1}
	20	0.826	0.737	0.827	0.860	0.135	0.052	0.737	0.112
	40	0.661	0.520	0.657	0.707	0.173	0.072	0.520	0.104
	120	0.390	0.238	0.374	0.422	0.181	0.087	0.237	0.049
	200	0.290	0.157	0.271	0.310	0.168	0.085	0.156	0.027
	300	0.227	0.112	0.207	0.239	0.154	0.080	0.111	0.014
	400	0.189	0.088	0.170	0.197	0.143	0.076	0.087	0.007
1.5	Linear K_ℓ ➤	3.456×10^{-2}	1.820×10^{-2}	1.95×10^{-1}	1.03×10^{-1}	1.95×10^{-1}	1.03×10^{-1}	8.92×10^{-2}	8.92×10^{-2}
	20	0.724	0.618	0.712	0.695	0.155	0.057	0.618	0.066
	40	0.535	0.406	0.517	0.491	0.169	0.066	0.405	0.054
	120	0.288	0.179	0.268	0.243	0.148	0.062	0.179	0.027
	200	0.210	0.119	0.191	0.172	0.131	0.056	0.119	0.017
	300	0.162	0.085	0.146	0.131	0.117	0.051	0.085	0.011
	400	0.135	0.067	0.120	0.108	0.108	0.047	0.067	0.007
2	Linear K_ℓ ➤	4.218×10^{-2}	1.236×10^{-2}	2.30×10^{-1}	9.70×10^{-2}	2.30×10^{-1}	9.70×10^{-2}	6.06×10^{-2}	6.06×10^{-2}
	20	0.631	0.577	0.617	0.584	0.148	0.047	0.577	0.041
	40	0.448	0.387	0.431	0.397	0.150	0.047	0.387	0.036
	120	0.235	0.181	0.221	0.202	0.125	0.037	0.181	0.022
	200	0.171	0.122	0.158	0.147	0.111	0.032	0.122	0.015
	300	0.132	0.088	0.121	0.113	0.100	0.029	0.088	0.011
	400	0.110	0.069	0.100	0.094	0.092	0.026	0.069	0.008

Table A39. Large-deflection Reduction Coefficients K_r

LOADING: Linearly varying pressure

$W = K_\ell K_r Q$, $\quad \bar{\sigma} = (t/a)^2 . E\bar{\sigma} \quad W = w/t$

$\bar{\sigma} = K_\ell K_r Q$, $\quad Q = a^4 q_{max}/t^4 E$

b/a	Q	W_1	$\bar{\sigma}_{bx1}$	$\bar{\sigma}_{by1}$	$\bar{\sigma}_{mx1}$	$\bar{\sigma}_{my1}$	$\bar{\sigma}_{by5}$	$\bar{\sigma}_{mx5}$	$\bar{\sigma}_{my5}$
0.67	Linear K_ℓ	3.874×10^{-3}	4.22×10^{-2}	6.41×10^{-2}	4.22×10^{-2}	6.41×10^{-2}	-1.51×10^{-1}	4.54×10^{-2}	1.51×10^{-1}
	20	0.999	0.999	0.999	0.015	0.010	1.000	-0.009	0.004
	40	0.997	0.994	0.996	0.030	0.019	0.998	-0.018	0.009
	120	0.965	0.943	0.953	0.085	0.054	0.978	-0.050	0.024
	200	0.917	0.867	0.891	0.125	0.078	0.947	-0.074	0.036
	300	0.854	0.770	0.809	0.159	0.098	0.905	-0.095	0.047
	400	0.793	0.682	0.733	0.178	0.108	0.864	-0.107	0.054
1	Linear K_ℓ	8.897×10^{-3}	8.55×10^{-2}	7.54×10^{-2}	8.55×10^{-2}	7.54×10^{-2}	-2.13×10^{-1}	6.38×10^{-2}	2.13×10^{-1}
	20	0.999	0.999	0.999	0.029	0.033	1.000	-0.033	0.009
	40	0.980	0.971	0.970	0.056	0.063	0.989	-0.064	0.017
	120	0.862	0.806	0.796	0.123	0.141	0.926	-0.149	0.040
	200	0.750	0.659	0.643	0.147	0.170	0.860	-0.188	0.053
	300	0.647	0.531	0.512	0.155	0.180	0.793	-0.208	0.061
	400	0.572	0.444	0.424	0.154	0.180	0.739	-0.215	0.066
1.5	Linear K_ℓ	1.319×10^{-2}	1.19×10^{-1}	6.14×10^{-2}	1.19×10^{-1}	6.14×10^{-2}	-2.53×10^{-1}	7.58×10^{-2}	2.53×10^{-1}
	20	0.989	0.986	0.981	0.033	0.051	0.995	-0.060	0.010
	40	0.954	0.941	0.923	0.060	0.094	0.980	-0.113	0.019
	120	0.778	0.721	0.647	0.106	0.165	0.898	-0.234	0.042
	200	0.652	0.576	0.482	0.112	0.176	0.830	-0.280	0.054
	300	0.550	0.465	0.369	0.109	0.173	0.765	-0.303	0.062
	400	0.482	0.394	0.303	0.104	0.167	0.713	-0.311	0.067
2	Linear K_ℓ	1.480×10^{-2}	1.30×10^{-1}	4.84×10^{-2}	1.30×10^{-1}	4.84×10^{-2}	-2.63×10^{-1}	7.90×10^{-2}	2.63×10^{-1}
	20	0.983	0.980	0.977	0.028	0.045	0.992	-0.079	0.011
	40	0.940	0.928	0.914	0.050	0.077	0.972	-0.145	0.021
	120	0.752	0.706	0.644	0.085	0.112	0.878	-0.286	0.048
	200	0.630	0.570	0.496	0.088	0.109	0.807	-0.342	0.062
	300	0.533	0.465	0.395	0.086	0.103	0.738	-0.371	0.073
	400	0.468	0.397	0.333	0.083	0.097	0.683	-0.381	0.080

Appendix **189**

Table A40. Large-deflection Reduction Coefficients K_r

LOADING : Linearly varying pressure

$W = K_\ell K_r Q$, $\bar{\sigma} = (t/a)^2 . E\bar{\sigma}$ $W = w/t$

$\bar{\sigma} = K_\ell K_r Q$, $Q = a^4 q_{max}/t^4 E$

b/a	Q	W_1	$\bar{\sigma}_{bx1}$	$\bar{\sigma}_{by1}$	$\bar{\sigma}_{mx1}$	$\bar{\sigma}_{my1}$	$\bar{\sigma}_{by5}$	$\bar{\sigma}_{mx5}$	$\bar{\sigma}_{my5}$
0.67	Linear K_ℓ ➤	3.874×10^{-3}	4.22×10^{-2}	6.41×10^{-2}	4.22×10^{-2}	6.41×10^{-2}	-1.51×10^{-1}	4.54×10^{-2}	1.51×10^{-2}
	20	1.000	1.000	1.000	0.023	0.010	1.000	-0.002	0.004
	40	0.994	0.992	0.993	0.045	0.019	0.996	-0.004	0.008
	120	0.956	0.934	0.945	0.125	0.053	0.971	-0.010	0.024
	200	0.898	0.847	0.872	0.182	0.075	0.931	-0.013	0.035
	300	0.839	0.764	0.800	0.237	0.095	0.891	-0.014	0.046
	400	0.756	0.651	0.701	0.253	0.098	0.832	-0.011	0.049
1	Linear K_ℓ ➤	8.897×10^{-3}	8.55×10^{-2}	7.54×10^{-2}	8.55×10^{-2}	7.54×10^{-2}	-2.13×10^{-1}	6.38×10^{-2}	2.13×10^{-1}
	20	0.992	0.990	0.990	0.048	0.033	0.996	-0.007	0.008
	40	0.965	0.953	0.953	0.090	0.061	0.980	-0.014	0.016
	120	0.804	0.743	0.741	0.183	0.122	0.882	-0.026	0.036
	200	0.675	0.586	0.581	0.211	0.137	0.797	-0.027	0.045
	300	0.567	0.462	0.455	0.219	0.139	0.719	-0.024	0.050
	400	0.493	0.382	0.374	0.218	0.136	0.662	-0.020	0.052
1.5	Linear K_ℓ ➤	1.319×10^{-2}	1.19×10^{-1}	6.15×10^{-2}	1.19×10^{-1}	6.15×10^{-2}	-2.53×10^{-1}	7.58×10^{-2}	2.53×10^{-1}
	20	0.975	0.968	0.963	0.063	0.050	0.987	-0.010	0.010
	40	0.910	0.889	0.870	0.109	0.085	0.954	-0.017	0.018
	120	0.667	0.602	0.550	0.172	0.124	0.815	-0.026	0.034
	200	0.530	0.450	0.392	0.177	0.121	0.722	-0.024	0.041
	300	0.429	0.346	0.290	0.173	0.112	0.644	-0.020	0.045
	400	0.366	0.284	0.231	0.166	0.103	0.589	-0.015	0.047
2	Linear K_ℓ ➤	1.480×10^{-2}	1.30×10^{-1}	4.84×10^{-2}	1.30×10^{-1}	4.84×10^{-2}	-2.63×10^{-1}	7.90×10^{-2}	2.63×10^{-1}
	20	0.960	0.952	0.945	0.067	0.043	0.980	-0.009	0.011
	40	0.874	0.851	0.832	0.111	0.066	0.934	-0.014	0.020
	120	0.609	0.549	0.508	0.160	0.077	0.772	-0.017	0.037
	200	0.477	0.407	0.366	0.163	0.070	0.672	-0.013	0.044
	300	0.384	0.312	0.276	0.158	0.062	0.591	-0.007	0.048
	400	0.326	0.255	0.223	0.152	0.056	0.534	-0.002	0.049

Thin Plate Design For Transverse Loading

Table A41. Large-deflection Reduction Coefficients K_r

LOADING : Linearly varying pressure

$W = K_\ell \, K_r \, Q$, $\quad \sigma = (t/a)^2 . E\bar{\sigma}$ $\quad W = w/t$

$\bar{\sigma} = K_\ell \, K_r \, Q$, $\quad Q = a^4 q_{max}/t^4 E$

b/a	Q	W_1	W_3	$\bar{\sigma}_{bx1}$	$\bar{\sigma}_{by1}$	$\bar{\sigma}_{mx1}$	$\bar{\sigma}_{my1}$	$\bar{\sigma}_{bx3}$	$\bar{\sigma}_{mx3}$
0.67	Linear K_ℓ	3.908×10^{-3}	4.290×10^{-3}	3.95×10^{-2}	4.09×10^{-2}	3.95×10^{-2}	4.09×10^{-2}	3.67×10^{-2}	3.67×10^{-2}
	20	0.999	0.999	0.999	0.999	0.008	0.010	0.999	-0.002
	40	0.996	0.994	0.995	0.996	0.017	0.019	0.994	-0.005
	120	0.969	0.955	0.962	0.972	0.048	0.055	0.953	-0.016
	200	0.927	0.896	0.908	0.930	0.074	0.084	0.891	-0.031
	300	0.867	0.819	0.831	0.867	0.099	0.110	0.810	-0.050
	400	0.809	0.750	0.758	0.804	0.115	0.127	0.738	-0.067
1	Linear K_ℓ	6.285×10^{-3}	2.388×10^{-3}	6.06×10^{-2}	4.89×10^{-2}	6.06×10^{-2}	4.89×10^{-2}	2.38×10^{-2}	2.38×10^{-2}
	20	0.997	1.001	0.996	0.995	0.018	0.024	1.001	-0.045
	40	0.989	0.998	0.985	0.982	0.036	0.046	1.002	-0.088
	120	0.925	1.031	0.892	0.874	0.089	0.116	1.013	-0.219
	200	0.846	1.059	0.785	0.749	0.117	0.153	1.026	-0.280
	300	0.761	1.079	0.674	0.621	0.131	0.173	1.033	-0.297
	400	0.693	1.080	0.590	0.528	0.135	0.180	1.025	-0.286
1.5	Linear K_ℓ	9.032×10^{-3}	0.5782×10^{-3}	8.18×10^{-2}	4.17×10^{-2}	8.18×10^{-2}	4.17×10^{-2}	0.794×10^{-2}	0.794×10^{-2}
	20	0.992	1.037	0.990	0.989	0.020	0.044	1.023	-0.250
	40	0.971	1.141	0.963	0.958	0.038	0.083	1.087	-0.478
	120	0.843	1.967	0.805	0.774	0.082	0.166	1.614	-1.058
	200	0.739	3.092	0.679	0.618	0.097	0.183	2.359	-1.244
	300	0.652	4.443	0.579	0.489	0.100	0.179	3.274	-1.137
	400	0.593	5.209	0.512	0.408	0.097	0.170	3.791	-0.885
2	Linear K_ℓ	1.016×10^{-2}	2.896×10^{-5}	8.95×10^{-2}	3.23×10^{-2}	8.95×10^{-2}	3.23×10^{-2}	0.187×10^{-2}	0.187×10^{-2}
	20	0.989	1.656	0.988	0.994	0.013	0.047	1.096	-1.638
	40	0.959	3.713	0.954	0.975	0.025	0.084	1.397	-3.039
	120	0.814	21.75	0.789	0.831	0.063	0.146	4.044	-6.195
	200	0.710	60.13	0.668	0.671	0.078	0.143	9.711	-7.109
	300	0.635	114.8	0.582	0.544	0.078	0.122	17.82	-5.776
	400	0.583	137.5	0.524	0.469	0.073	0.108	21.15	-3.840

Appendix 191

Table A42. Large-deflection Reduction Coefficients K_r

LOADING: Linearly varying pressure

$W = K_\ell K_r Q$, $\quad \sigma = (t/a)^2 \cdot E\bar{\sigma}$, $\quad W = w/t$

$\bar{\sigma} = K_\ell K_r Q$, $\quad Q = a^4 q_{max}/t^4 E$

b/a	Q	W_1	W_3	$\bar{\sigma}_{bx1}$	$\bar{\sigma}_{by1}$	$\bar{\sigma}_{mx1}$	$\bar{\sigma}_{my1}$	$\bar{\sigma}_{bx3}$	$\bar{\sigma}_{mx3}$
0.67	Linear K_ℓ ➡	3.908×10^{-3}	4.291×10^{-3}	3.95×10^{-2}	4.09×10^{-2}	3.95×10^{-2}	4.09×10^{-2}	3.68×10^{-2}	3.68×10^{-2}
	20	0.998	0.996	0.997	0.999	0.022	0.010	0.996	0.012
	40	0.991	0.985	0.990	0.994	0.043	0.019	0.984	0.024
	120	0.934	0.892	0.925	0.954	0.116	0.052	0.886	0.053
	200	0.862	0.781	0.840	0.897	0.165	0.075	0.769	0.056
	300	0.779	0.664	0.742	0.823	0.201	0.093	0.646	0.046
	400	0.710	0.575	0.661	0.756	0.222	0.105	0.554	0.031
1	Linear K_ℓ ➡	6.285×10^{-3}	2.388×10^{-3}	6.06×10^{-2}	4.89×10^{-2}	6.06×10^{-2}	4.89×10^{-2}	2.38×10^{-2}	2.38×10^{-2}
	20	0.996	0.994	0.995	0.995	0.033	0.024	0.994	-0.007
	40	0.978	0.971	0.972	0.972	0.064	0.045	0.970	-0.014
	120	0.868	0.831	0.830	0.832	0.149	0.106	0.825	-0.036
	200	0.755	0.698	0.691	0.693	0.185	0.132	0.687	-0.049
	300	0.649	0.582	0.565	0.568	0.202	0.144	0.565	-0.057
	400	0.571	0.501	0.478	0.481	0.207	0.147	0.481	-0.062
1.5	Linear K_ℓ ➡	9.032×10^{-3}	5.782×10^{-4}	8.18×10^{-2}	4.17×10^{-2}	8.18×10^{-2}	4.17×10^{-2}	0.794×10^{-2}	0.794×10^{-2}
	20	0.984	0.992	0.981	0.979	0.043	0.044	0.989	-0.006
	40	0.945	0.971	0.932	0.927	0.079	0.080	0.962	-0.010
	120	0.750	0.853	0.701	0.684	0.148	0.141	0.816	-0.019
	200	0.615	0.753	0.549	0.529	0.164	0.150	0.702	-0.022
	300	0.509	0.660	0.434	0.415	0.168	0.147	0.603	-0.024
	400	0.439	0.590	0.362	0.344	0.166	0.141	0.532	-0.026
2	Linear K_ℓ ➡	1.016×10^{-2}	2.896×10^{-5}	8.95×10^{-2}	3.23×10^{-2}	8.95×10^{-2}	3.23×10^{-2}	0.186×10^{-2}	0.186×10^{-2}
	20	0.974	1.120	0.970	0.974	0.046	0.045	1.002	0.028
	40	0.915	1.376	0.902	0.915	0.081	0.077	1.002	0.049
	120	0.688	2.085	0.646	0.678	0.139	0.115	0.952	0.076
	200	0.556	2.244	0.501	0.535	0.152	0.116	0.876	0.074
	300	0.458	2.203	0.396	0.426	0.155	0.110	0.789	0.066
	400	0.394	2.094	0.330	0.357	0.153	0.103	0.717	0.058

Index

Accuracy 17, 65
Aircraft wing 30
Airy's stress function 12
Aluminium plate 111
Aspect ratio 64, 113, 147

Behaviour
  large-deflection 2, 13, 136
  membrane 12
  small-deflection 2, 11, 140
Bibliography 123–128
Boundary conditions
  description 2, 19–31, 63
  equations 14–16
  idealisations 23–31
  symbolic representation 21, 131
Bridges, steel deck 27

Concentrated loading 38, 69–84, 117
Container, liquid 25, 119
Criteria
  design 31–36
  yield 33

Data 44–108, 151–191
Deflections
  central 38
  large 38
  small 38
Design
  criteria 31–36
  curves 44–108
  examples 24, 26, 27, 30, 109–122, 142–143
  large-deflection 18–36
  parameters 37
  procedure 38
  small-deflection 38, 134
  tables 150–191
Distribution, stress 56–63

Edge beam 25, 94, 103–104, 119
Equations
  boundary conditions 14–16
  equivalent stress 33
  flexural 11, 12
  large-deflection 14

194 Index

maximum stress 38
membrane 13, 14
Equivalent stress 33, 38, 59, 141
Examples
aircraft 30, 143
boundary conditions 23–31, 145
bridges 27
containers 25
glass 24
numerical 109–122, 142
ships 27

Flexural actions 7
Floating platform 27
Fracture 32

Geometry 18
Glass 24, 115, 147

Historical background 5
Hydrostatic pressure 94–102, 119

Large-deflection
behaviour 2, 13, 136
coefficients 136
design 18–36
equations 14
stresses 137
tables 151–191
List of cases treated 41–43, 132–134
Loading 2, 18, 37, 38, 132
concentrated 38, 69–84, 117

Membrane
actions 7
behaviour 12
forces 13
Moments 12

Plasticity 33–36, 121
Plate
behaviour 1, 2, 11, 13, 136, 140
circular 43, 104–108
elements of 7
rectangular 41–104

Poisson's ratio 40, 136, 141, 147
adjustment for 40
Pressure
hydrostatic 2, 26, 37, 94–102, 119
uniform 2, 37

Reduction coefficients 134–136
References 123–128

Sectional actions 6
Shearing stress 139
Ships, plates 27
Signs 8, 9, 139
Small-deflection
behaviour 2, 11, 140
coefficients 134, 140
equations 14
stresses 137
Solution, computer 17
Solutions, approximate 64–69
Steel sheet 23, 109, 117, 143
Strain 12, 13
Stress
bending 10, 56–59
distribution 56–63
equations 12
equivalent 33, 38, 59, 141
function 12
large-deflection 10, 137, 141
maximum tensile 38, 59
membrane 10, 56, 57
shear 139
surface 10, 141
total 10

Transverse pressure 2, 37

Von Mises 33

Wheel loading 23, 38, 69–84, 117

Yield
criteria 33
mid-plane 35
thickness penetration 34, 121